Windscale 1957

Also by Lorna Arnold
BRITAIN AND THE H-BOMB

BRITAIN, AUSTRALIA AND THE BOMB

Windscale 1957

Anatomy of a Nuclear Accident

Third Edition

Lorna Arnold
Official Historian

© The United Kingdom Atomic Energy Authority 1992, 1995, 2007
Foreword to the Third Edition(1) © Peter Hennessy 2007
Foreword and note on p. 231 to the Third Edition(2) © Brian Cathcart 2007
Foreword to the First Edition © Sir Alan Cottrell 1992, 1995, 2007

All rights reserved. No reproduction, copy or transmission of this publication may be made without written permission.

No paragraph of this publication may be reproduced, copied or transmitted save with written permission or in accordance with the provisions of the Copyright, Designs and Patents Act 1988, or under the terms of any licence permitting limited copying issued by the Copyright Licensing Agency, 90 Tottenham Court Road, London W1T 4LP.

Any person who does any unauthorized act in relation to this publication may be liable to criminal prosecution and civil claims for damages.

The author has asserted her right to be identified as the author of this work in accordance with the Copyright, Designs and Patents Act 1988.

First published 1992 by Macmillan Press Ltd
Second edition published 1995 by Macmillan Press Ltd
This edition published 2007 by
PALGRAVE MACMILLAN
Houndmills, Basingstoke, Hampshire RG21 6XS and
175 Fifth Avenue, New York, N.Y. 10010
Companies and representatives throughout the world

PALGRAVE MACMILLAN is the global academic imprint of the Palgrave Macmillan division of St. Martin's Press, LLC and of Palgrave Macmillan Ltd. Macmillan® is a registered trademark in the United States, United Kingdom and other countries. Palgrave is a registered trademark in the European Union and other countries.

ISBN-13: 978-0-230-57317-8   paperback
ISBN-10: 0-230-57317-7   paperback

This book is printed on paper suitable for recycling and made from fully managed and sustained forest sources. Logging, pulping and manufacturing processes are expected to conform to the environmental regulations of the country of origin.

A catalogue record for this book is available from the British Library.

A catalog record for this book is available from the Library of Congress.

10 9 8 7 6 5 4 3 2 1
16 15 14 13 12 11 10 09 08 07

Printed and bound in Great Britain by
Antony Rowe Ltd, Chippenham and Eastbourne

Contents

List of Appendices	vi
Foreword to the Third Edition (1) by Peter Hennessy	vii
Foreword to the Third Edition (2) by Brian Cathcart	ix
Introduction to the Second Edition	xi
Foreword to the First Edition by Sir Alan Cottrell, FRS	xii
Note on Documentation	xiii
List of Abbreviations	xiv
Glossary	xvi
Introduction	xxi
1 Britain's Atomic Bomb	1
2 Windscale's Origins	8
3 After *Hurricane*	19
4 The Ninth Anneal	42
5 Damage Assessment and Damage Control	60
6 The Penney Inquiry and the First White Paper	77
7 Three More White Papers	98
8 Causes: An Accident Waiting to Happen	124
9 Appraisals and Reappraisals	136
10 Postscript	154
Appendices	161
Note on Sources	205
Notes and References	208
Bibliography	225
Developments since 1995	231
Index	232

List of Appendices

I	Chronology of events, October 1957–October 1958	161
II	Responsibilities and organisation of the IG	164
III	Instruction of 14 November 1955 on Wigner releases	167
IV	Summary of Wigner releases in Windscale piles, 1953–57	168
V	Note on uranium fuel cartridges in the Windscale piles	169
VI	Note on other cartridges in the Windscale piles	172
VII	Emergency site procedure at Windscale	175
VIII	Calculations of emergency levels for iodine-131	178
IX	Estimates of fission product and other radioactive releases resulting from the 1957 fire in Windscale Pile No.1	184
X	Estimates of total radiological impact of the radioactive releases resulting from the 1957 fire in Windscale Pile No.1	187
XI	Report of Penney Inquiry	189

Foreword to the Third Edition (1)

In the 1950s, I had family in Cumberland. They lived in St. Bees. My Uncle George (a bit of a hero because he had been on the 1933 Everest expedition) had worked as an engineer building Calder Hall until his early death in the mid-Fifties. As a young boy, in the years before the Windscale fire, I had visited St. Bees and Coniston, Hawkshead and Ambleside. That very summer of 1957 I had been in Borrowdale. The Lake District was a vivid, pleasurable part of my formation.

In October 1957, therefore, I watched avidly the Movietone depiction of the fire at my local cinema in north London. I have a clear memory of milk pouring from churns into drains in Cumberland and Westmorland for fear that it had become contaminated with radioactive iodine. Little did I realise how close we came to the Lake District being sealed off for a good part (if not all) of my remaining lifetime. I shall never forget Lord Plowden in the 1980s reliving for me the decision which had to be taken when he was Chairman of the Atomic Energy Authority in the autumn of 1957 about whether or not to douse Windscale pile no. 1 with water.

As Lorna Arnold puts it in this fine book: 'Water was a last resort. Putting water on burning graphite and metal might cause an explosive mixture of carbon monoxide and hydrogen with air...' No-one could predict what would happen as the fire raged at 400 degrees centigrade. Had the pile exploded, those incomparable fells would have been turned in a flash into a hazardous wasteland.

But, as Lorna puts it, on 11 October 1957, water did douse the fire 'and an environmental catastrophe had been averted'. Professional historians are meant to eschew counter-factual, 'what if' speculation. In the case of the Windscale fire, it is irresistible.

Had it exploded, the repercussions would have been immense not just for those living and working in the shadow of the deadly debris. 1957 would be remembered globally for two things: the Russian 'Sputnik' (which was in space just a few days before the graphite caught alight) and the Windscale fire. Not only would the possibility of renewed nuclear weapons collaboration between the USA and the UK have been halted in its tracks, so, very probably, would have been the still pioneering civil nuclear programmes in the UK, France, possibly the USA and even the Soviet Union.

Mercifully, 1957 is not so remembered. But this great near-miss of a story really does need rescuing from history for the enlightenment of a new generation – not least because a new era of civil nuclear power station construction is upon us, as is an upgrading of the Royal Navy's Trident strategic nuclear deterrent. Nobody is better placed than the incomparable Lorna Arnold to relive and to tell this timely story. She does it magnificently in *Windscale 1957*.

PETER HENNESSY, FBA
Attlee Professor of Contemporary British History
Queen Mary, University of London, UK

Foreword to the Third Edition (2)

This new edition of the history of the Windscale accident appears fifty years after the events it describes; neatly, it happens also to be forty years since the book's author, Lorna Arnold, began her career as official historian of the British nuclear project. The book itself, and her other works, are evidence enough of her contribution to the understanding of a complex and controversial field, but it would be wrong to allow this occasion to pass without adding a little to the record about the remarkable writer herself.

It might surprise the reader to learn that at the time this book was first published in 1992 Lorna Arnold was aged seventy-six, that she would publish a further important book, on Britain's H-bomb, nine years later, and (with a co-author, Mark Smith) a substantially revised version of an earlier work on nuclear tests in Australia five years after that. Indeed she is still writing in 2007, at the age of ninety-one.

But longevity is the least of her distinctions, for she embodies all that is best in official history, a function that is rarely valued as it should be. She has worked from the inside, as it were, of the nuclear establishment, with privileged access to records, people and expertise, but the portraits of events that she has painted over four decades have always been warts-and-all ones. Those who know her will testify that she is by nature a sceptic when it comes to authority, at times even a dissident, and the eye she casts on the nuclear world has never been an indulgent one. It is her fairness and humanity that have enabled her to function so well in, and reveal so much about, a sphere which, more than almost any other, has a deeply-rooted aversion to disclosure. A whole generation of scientists, engineers and administrators have trusted her, and they have been rewarded with straight treatment.

That she ever entered this particular field owes a good deal to the Windscale fire. As she explains in the pages that follow, the accident prompted a transformation of the British atomic establishment that included a rapid increase in staff numbers. At the time, in 1958, she was a single mother of two boys, working at a job she hated. The UK Atomic Energy Authority brought her in to be secretary to a committee on training in radiological health and safety. Nine years later she became a historian at the Authority, joining Margaret Gowing to help produce the monumental *Independence and Deterrence: Britain and Atomic Energy 1945–52*, published in 1974. The plan after that was to write a successor work on a similar scale covering the years to 1958, but though much work was done it never came to fruition. Instead, Lorna Arnold has produced three valuable books of her own which serve for historians as the foundations for the study of the nuclear project in the 1950s.

Besides its fairness and its independent stance, her work – and this book in particular – is marked by other qualities. There is her refusal to be intimidated by the science, and there is her gift for incorporating the human element alongside the documented record. The latter is usually the fruit of many interviews and often of long acquaintance with her subjects. There is also her distaste for anachronism: she will not impose the values of one generation upon an earlier one. And a further quality deserves note, which will be obvious when you reach Chapter 10 if it has not become plain long before: Lorna Arnold is a very calm and wise reader of events.

BRIAN CATHCART
Professor of Journalism, Kingston University,
London, UK

Introduction to the Second Edition

It is more than forty years since the Windscale piles began producing plutonium for Britain's first atomic bombs, and over thirty-five years since both piles were closed after a serious fire in Pile No. 1 – the world's first major reactor accident. But the story of the piles will not be finished until they have been dismantled and safely disposed of.

After the 1957 accident as much fuel as possible was discharged from the piles; then they were left, monitored and regularly inspected but undisturbed. By allowing time for decay of radioactivity of short and medium half-life, eventual decommissioning would become somewhat easier and safer. However some work became necessary in 1987 when a structural survey showed defects in the chimneys, and a programme of work to remove the filter galleries and to cap the remaining structures was instituted and is still under way.

In 1992 the Advisory Committee on the Safety of Nuclear Installations (ACSNI) carried out a survey of radioactive waste management and decommissioning in the UKAEA. Reporting in March 1994, it expressed concern about the piles which, it said, fell 'far below the "as safe as reasonably achievable" criterion'. It foresaw a substantial programme of work to get them into a safe condition, including removal from Pile No. 1 of the damaged fuel elements – perhaps twenty tons altogether. When the ACSNI report appeared some preliminary work had already been done, further work was being planned, and W. S. Atkins (Northern) had been appointed to oversee it on behalf of the UKAEA.

To meet the remote possibility of a significant earthquake in Cumbria the structures are being sealed, and seismically qualified ventilation plant is also being provided. Future decommissioning options, including total dismantling, are still being assessed. So the old Windscale piles, though of unique and obsolete design – 'monuments to our initial ignorance' as Lord Hinton called them – are still important; they are not only of historical interest. Their final decommissioning and disposal presents novel technological problems and is demanding and expensive. They will continue to be of interest – locally, to safety experts, to the nuclear industry, to environmentalists and to the public – for many years.

Foreword to the First Edition

Lest we forget is the historian's motivator and justification; to capture the facts and record them coherently and fairly before they escape the grasp of fading memories and perishable documents, or are supplanted by myth and wishful fiction. In the historian's traditional theatres – politics, military struggle, national leadership, imperialism, social evolution, ecclesiastical power – there is no shortage of researchers, interpreters, presenters and commentators. But the twentieth century has seen the rise of other great human activities, notably in science and technology, and the historians of these are still rare; a handful of pioneers struggling to ensure that these new branches of history do not suffer such nebulous and fabulous beginnings as the classical ones. That is why we must be grateful to Lorna Arnold and her colleagues, pre-eminently Margaret Gowing, who are endeavouring to capture our fading technological history.

The Windscale accident of 1957 is the equivalent of a wartime battle. All the same basic elements are there: misjudgements, professional rivalries, brilliant improvisation, desperate decisions and heroic actions, all wrapped in a cloud of uncertainty as dense as any fog of war. As a beautiful account of epic endeavour, exposing human character in all its complexity as realistically as in any battle, Lorna Arnold's story grips the reader by its sheer fascination; a true adventure, expertly told by a professional historian.

SIR ALAN COTTRELL

Note on Documentation

In accordance with the practice of official histories commissioned by the Cabinet Office, references to official papers that are not yet publicly available have been omitted; notes are confined to published material or documents in the Public Record Office.

The author and publishers are grateful to the Controller of Her Majesty's Stationery Office for permission to reproduce and quote Government documents. She also acknowledges her indebtedness to the authors and publishers of all those books that have been mentioned in this work.

List of Abbreviations

AEA	Atomic Energy Authority. Used to refer to the AEA Board, or the whole organisation (see also UKAEA)
AERE	Atomic Energy Research Establishment (Harwell)
AEX	Atomic Energy Executive (consisting of the full-time AEA Board Members)
AGR	Advanced Gas-Cooled Reactor
AHSB	Authority Health and Safety Branch
AM	Code-name for tritium, hence applied to the cartridges containing rods of lithium-magnesium (Li-Mg) alloy irradiated in the piles to produce tritium
ARC	Agricultural Research Council
AWRE	Atomic Weapons Research Establishment (Aldermaston); transferred from the AEA to the Ministry of Defence in 1973, now AWE
BCDG	Burst Cartridge Detection Gear, used to detect and locate faulty fuel elements in the piles so that they could be discharged before causing trouble
BEPO	Experimental low power, air-cooled, graphite-moderated reactor at Harwell, commissioned in 1948
BNFL	British Nuclear Fuels Limited, created in 1971 by hiving off the Production Group from the AEA (now British Nuclear Fuels plc)
CEA	Commissariat à l'Energie Atomique (France)
CEA	Central Electricity Authority (the predecessor of the CEGB)
CEGB	Central Electricity Generating Board, set up in 1957
COMARE	Committee on Medical Aspects of Radiation in the Environment
CSAR	Chief Superintendent of Armament Research
DFR	Dounreay Fast Reactor, a 60 megawatt experimental reactor at Dounreay, in the north of Scotland; commissioned in 1959
ENEA	European Nuclear Energy Agency
GLEEP	Low energy research reactor, air-cooled and graphite-moderated, at Harwell; commissioned in 1947
IAEA	International Atomic Energy Agency

List of Abbreviations   xv

ICI	Imperial Chemical Industries
ICRP	International Commission on Radiological Protection
IG	The Industrial Group of the AEA, with headquarters at Risley, near Warrington, Lancashire (now Cheshire)
IGY	International Geophysical Year
IPCS	Institution of Professional Civil Servants
LM	Code-name for polonium-210; also applied to the cartridges containing bismuth oxide which were irradiated in the piles to produce polonium-210
MAFF	Ministry of Agriculture, Fisheries and Food
MHLG	Ministry of Housing and Local Government
MRC	Medical Research Council
NATO	North Atlantic Treaty Organization
NDA	Nuclear Decommissioning Authority
NII	Nuclear Installations Inspectorate
NRPB	National Radiological Protection Board
PEC	Production Executive Committee
PERG	Political Ecology Research Group
PIPPA	A type of dual-purpose, gas-cooled, graphite-moderated reactor, designed to produce weapons-grade plutonium with electricity as a by-product
PIRC	MRC Committee on Protection against Ionising Radiations
PRO	Public Record Office
QFE	Quartz fibre electrometer
R & D	Research and development
R & DB	Research and Development Branch (Industrial Group)
R & DB (W)	Research and Development Branch, Windscale
RPD	Radiological Protection Division of the Authority Health and Safety Branch, located at Harwell
SRD	Safety and Reliability Directorate
SU	Sunshine units (used for strontium-90)
TEC	Technical Evaluation Committee (the Fleck Committee)
TX	Technical Executive Committee, set up by the AEA's Industrial Group immediately after the 1957 Windscale accident
UKAEA	United Kingdom Atomic Energy Authority, set up in 1954 (see also AEA)
UNSCEAR	United Nations Scientific Committee on Effects of Atomic Radiation
USAEC	United States Atomic Energy Commission (1946–74)

Glossary

Alpha radiation See Ionising radiation.

Atomic bomb (A-bomb) Bomb in which the explosive power is derived from the fission of plutonium or uranium-235.

Becquerel (Bq) A measure of radioactivity, equivalent to the activity of a radionuclide that decays at the rate of one nuclear distintegration per second (1 d.p.s.). It is named after Henri Becquerel who discovered the radioactivity of uranium. It replaced the curie (Ci), a unit based on the radioactivity of 1g of radium, equivalent to 3.7×10^{10} d.p.s.: $1Bq = 27 \times 10^{-12} Ci$.

Beta radiation See Ionising radiation.

Boron An element which strongly absorbs slow neutrons, which is therefore used in steel alloys for making reactor control rods. It is also used for hardening steel.

Caesium A silvery-white metal (in its non-radioactive form widely used in the electronics industry). Radioactive caesium-137 is a fission product with a half-life of 30 years, which emits beta radiation.

Collective dose equivalent The measure of the total radiation exposure of a group of people or a population.

Criticality The state in a nuclear reactor when the fission process becomes self-sustaining.

Critical size The minimum size for a reactor core of a given configuration.

Curie (Ci) A measurement of radioactivity (named after Marie and Pierre Curie) based on the activity of 1g of radium, 3.7×10^{10} disintegrations per second. It has been replaced by the becquerel (Bq: see above): $1Ci = 3.7 \times 10^{10} Bq$.

Divergence The initiation of a chain reaction in a reactor as the control rods are withdrawn.

Dosimeter Instrument for measuring radiation dose; commonly a pocket electroscope, or a small photographic film worn on the torso, wrist or finger and sometimes partly covered by cadmium and tin screens so that exposure to different types of radiation can be estimated separately.

Glossary xvii

Eutectic A mixture of two or more solid substances at a minimum melting point.

Film badge See Dosimeter.

Fissile material Heavy isotopes (uranium-235 and plutonium-239) which are capable of fission when they capture neutrons of suitable energy.

Fission The spontaneous or induced disintegration of heavy atoms into two or more lighter ones. The process involves a loss of mass which is converted into nuclear energy.

Fission products Atoms and particles, often radioactive, resulting from nuclear fission (e.g., strontium-90, iodine-131, and caesium-137), which are all important contributors to fallout.

Gray (Gy) Unit of radiation exposure (named after a British radiobiologist) which replaced the rad (see below): 1Gy = 100 rads.

Half-life The time in which half the atoms of a given quantity of a radionuclide undergo disintegration. Half-lives of different radionuclides can vary from fractions of a second to thousands or millions of years.

Health physics Scientific discipline concerned with health hazards of ionising radiations and protection measures required.

Hydrogen bomb (H-bomb) Bomb in which the explosive power is mainly derived from the fusion of light elements (deuterium or tritium) rather than the fission of the heavy elements uranium and plutonium.

Iodine-131 A volatile radioactive fission product with a half-life of 8 days, which emits beta and gamma radiation. If it enters the body it becomes concentrated in the thyroid gland.

Ionising radiation Radiation from man-made sources (e.g., medical X-rays, reactors, fallout) or natural sources (e.g., cosmic rays and terrestrial radiation from rocks and soil) which is able to ionise atoms in biological molecules, with various possible health effects. These include radiation sickness (in the case of acute exposure) and long-term ill effects – such as leukaemia or other cancers – in the case of single or repeated low exposures.

Ionising radiation, types of
Particulate radiation
(1) Alpha particles: these relatively heavy, positively charged, particles are the nuclei of helium atoms. They have so little penetrating power that they can be stopped by a sheet of paper; they are only a health hazard if

alpha-emitters (e.g., plutonium) are taken into the body, but can then cause extremely localised but very damaging irradiation of the tissues.

(2) Beta particles: these are electrons (negatively charged) and are very light. They are more penetrating than alpha particles. From an external source they reach only the surface tissues of the body, and are a hazard to the skin and to immediately underlying tissues, but not to deep tissues. But they are more hazardous if beta-emitters are taken into the body (e.g., by inhalation or ingestion).

(3) Neutrons: these uncharged particles are about one-quarter of the mass of an alpha particle. They are penetrating and can be very damaging biologically, but the circumstances in which people are likely to be exposed to neutron radiation are rare.

Electromagnetic radiation

(4) Gamma radiation: this consists of electromagnetic rays originating in the atomic nucleus; they are extremely penetrating and can affect the whole body, including the internal organs, when the source is external to the body. They are less effective when their parent material is inside the body.

(5) X-rays: these are similar to, and behave like, gamma rays.

Ionising radiation, measurement of exposure The units of radiation in use in 1957 were roentgen, rad and rem; now gray and sievert (*q.v.*) are used.

Isotope The form of an element. Most elements occur in several forms which are chemically identical (with the same atomic number) but which differ in atomic mass. The atomic number is identified with the number of protons (or positively charged particles) in the nucleus. The atomic mass number depends on how many nucleons (protons plus neutrons) there are in the nucleus. The number of neutrons (but not protons) varies in the different isotopes of an element. Unstable isotopes (radioisotopes) tend to disintegrate and emit radiation.

Lithium (Li) A light element chemically resembling sodium. Canned rods of lithium-magnesium alloy were irradiated in the Windscale piles to produce tritium (code-named AM).

Magnesium (Mg) A light metallic element, brilliantly white, which readily burns in air giving an intense white light. Used in strong light alloys.

Magnox A magnesium alloy used for canning fuel elements in the gas-cooled graphite-moderated reactors at Calder Hall and Chapelcross and

Glossary xix

the first generation of British civil nuclear power stations. The term is also applied to the reactor type itself.

Nuclide   Isotope of an element; there are some 1400 nuclides among the known elements.

Plutonium (Pu)   A man-made element of atomic number 94 (i.e., with 94 protons in its nucleus), therefore originally referred to as '94'. Plutonium-239 is created by neutron absorption when uranium-238 is irradiated in a nuclear reactor. The isotope has 94 protons and 145 neutrons – totalling 239 nucleons – in its nucleus, and is fissile. It is used in atomic weapons, and can also be used as a nuclear reactor fuel.

Polonium (Po)   A rare radioactive metal that occurs in some uranium ores. Polonium-210 (code-named LM), formerly used as an A-bomb component, was produced by irradiating cartridges containing bismuth oxide in a reactor.

Quality factor (Q)   Relative biological effectiveness of different types of radiation (Q = 1 for gamma radiation, Q = 20 for alpha radiation.)

Rad   A unit of absorbed radiation dose, now replaced by the gray (Gy): 1 rad = 0.01Gy.

Radiation   See Ionising radiation.

Radioiodine   Radioactive forms of iodine, especially the radioisotope iodine-131.

Radioisotopes   Unstable isotopes.

Radionuclide   Any nuclide (or form of an element) which is unstable and undergoes radioactive decay.

Rem   A unit of absorbed radiation dose usually equivalent to the rad (see above) but modified by a quality factor (Q) reflecting the capacity for biological damage of different types of ionising radiation. The rem has now been replaced by the sievert (Sv): 1 rem = 0.01Sv.

Safeguards   The assessment of the safety of reactors and plants (including laboratories), the preparation of safety manuals, and related theoretical and experimental studies.

Sievert (Sv)   A unit of absorbed radiation dose (named after a Swedish physicist) which has replaced the rem (see above): 1Sv = 100 rem.

Strontium (Sr) A silvery-white metallic element, chemically similar to calcium. Strontium-90, a beta emitter, is an important fission product with a half-life of 28 years. If it enters the body it concentrates in bone.

Thermo-nuclear bomb See Hydrogen bomb.

Tritium (T or H_3) A radioisotope of hydrogen, with a half-life of 12.5 years. It is an ingredient of thermonuclear weapons. It was produced at Windscale by irradiating cartridges containing rods of lithium-magnesium alloy (see above). Since the codename for tritium was AM, these cartridges were referred to as AM cartridges.

Uranium (U) A heavy, hard, grey metal with several isotopes of which uranium-238 is the most abundant. The fissile radioisotope uranium-235 exists only as one part in 140 in natural uranium. Natural uranium (or uranium slightly enriched in U-235) is used as fuel in reactors; uranium-235 is used as a nuclear explosive in some types of nuclear weapons.

Wigner effect The process by which an atom is knocked out of its position in a crystal lattice by nuclear impact. A change of chemical or physical properties may result (e.g., graphite bombarded by neutrons changes in dimension and shape).

Wigner energy Energy stored within the crystal lattice of, for example, graphite as a result of the Wigner effect.

Introduction

In November 1957 seventeen miners were killed and eleven injured in a mining disaster at Kames colliery, in Ayrshire. A month earlier, on the Cumbrian coast, there had been a fire in a nuclear reactor. Fortunately there were no deaths and no known injuries; local milk, as a safety precaution, was withdrawn from sale or consumption for a few weeks and was poured away; the farmers were fully compensated at the cost of several thousand pounds, and the ditches, it was said, stank of sour milk.

Most people soon forgot the Kames tragedy, but the Windscale accident captured the headlines, preoccupied Ministers – especially the Prime Minister – and engaged senior officials, scientists and engineers for weeks and months. It is still a live issue, and is still vividly – if often inaccurately – remembered. A folklore has grown up around it, with tales of clouds of black smoke, the slaughter of radioactive cattle and how the famous filters ('Cockcroft's follies') saved the day. In the United States there were even rumours that the accident had made the region uninhabitable for 200 years.[1]

Why this contradiction? Atomic energy had a special status. It was new and unfamiliar; even, to the public, mysterious. It was, moreover, a matter of high policy, involving international relations, the defence of the realm, and national prestige. By contrast, coal mining was an old and familiar industry, not an exciting advanced technology. Mining accidents were not infrequent and when explosions, roof falls or mechanical failures happened they were shocking but not incomprehensible. Their consequences, though they might be grievous for families and communities, were local. They were easily understood and the public did not feel menaced by them.

The 1957 Windscale fire went to the very heart of Britain's defence programme. It also threatened a new and glamorous civil technology. Few people in the country understood much about it, but nuclear power seemed brilliantly promising and full of hope: in Churchill's words 'a perennial fountain of world prosperity'.[2] And Britain proudly led the world; in October 1956, just a year before the fire, Queen Elizabeth II had ceremonially opened the first reactor at Calder Hall and for the first time anywhere nuclear-generated electricity flowed into a national grid. Besides building more reactors at Calder Hall – and at Chapelcross in Dumfriesshire – to produce plutonium and other atomic weapon materials, Britain in 1955 had launched the world's first civil nuclear power programme. It was an ambitious 1500–2000 megawatt programme of up to twelve twin-reactor stations, to be built in ten years; for a time the plan was trebled, to 5000–6000 megawatts.[3] At the same time Britain was creating an independent nuclear deterrent: testing, manufacturing and stockpiling atomic weapons and developing H-bombs. The military and civil nuclear programmes were both of supreme importance to the British government and the Windscale accident had an immediate impact on both.

The Windscale fire, again unlike the Kames disaster, had no obvious precedent. It was the first major and well publicised reactor accident with off-site consequences: another, but inauspicious, first. Unlike Kames, no one was perceptibly harmed. But it was serious; potentially (but not actually) a disaster, so that it could be – and was – generally referred to as an 'incident', or even a 'mishap'.

Nuclear experts in the United States and in Britain had given a good deal of thought, well before the Windscale fire, to the possible consequences of an off-site release of radioactivity from a reactor in the event of a serious accident. Technologies learn from mistakes and accidents, and nuclear technology is no exception, though it has been said that it could not afford accidents and must learn without them. Fortunately, such accidents have been rare. Three have been of world-wide public interest: Windscale, in 1957; Three Mile Island, in Pennsylvania, in 1979; and Chernobyl in the Ukraine, in 1986, by over a thousand times the biggest.

The aim of this book is to set the 1957 Windscale accident in its historical context in the immediate post-war period and the early days of the Cold War; to describe the event and its consequences; and to evaluate it from the vantage point of 1990. I have used official papers in several Government offices as well as in the Atomic Energy Authority (AEA) archives; all the extensive AEA documentation, with the exception of one or two items, was opened in the Public Record Office (PRO) in January 1988 and 1989, some of it in advance of the 30-year rule.

Interviews, conversations and correspondence have been most helpful in supplementing contemporary documents and elucidating technical problems and I am very grateful to all the people who, over the years, have kindly given their time to talk or write to me: particularly to Mr A. M. Allen, Mr H. J. Dunster, Dr Arthur Chamberlain, the late Mr W. Crone, Professor F. R. Farmer, Mr R. Gausden, Mr V. Goodwin, Dr G. B. Greenough, Sir John Hill, the late Lord Hinton, Mr H. Howells, Mr T. C. Hughes, Mr G. D. Ireland, Mr D. R. Mackey, the late Dr A. S. McLean, Mr J. S. Nairn, Lord Penney, Lord Plowden, Mr R. I. Robertson, Mr K. Saddington and Mr T. Tuohy.

My thanks and appreciation are due to the records staff in government departments and in the AEA, and especially to Mrs Vivienne Martin (now retired) for her enthusiastic help and outstanding professional skill as a departmental record officer, and to Miss Lynn Christer, Mrs Valerie Marson and Mrs Amy Singh, who typed my drafts so admirably. Finally my warmest thanks to Professor Margaret Gowing, my admired friend and colleague for over twenty years.

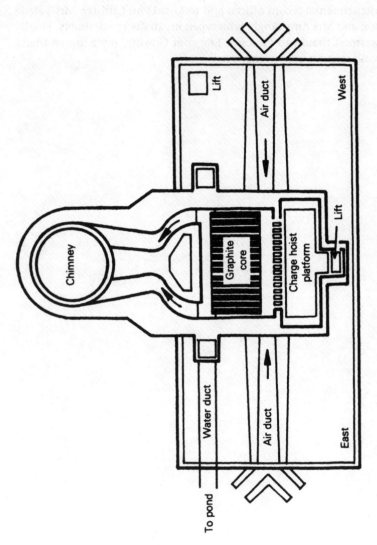

Figure 1 A plan of the pile

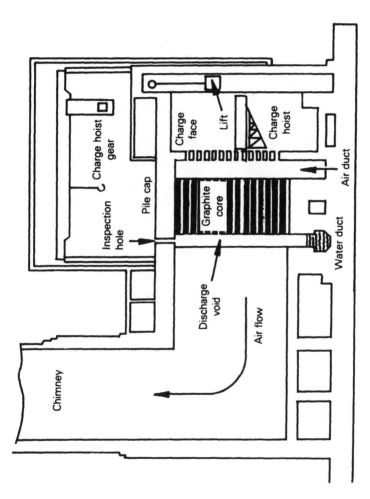

Figure 2 A section through the pile

1 Britain's Atomic Bomb

To understand the 1957 Windscale accident one must go back ten years and more to the genesis of the British atomic programme in the mid-1940s and the decision to make atomic weapons. The origins of that decision are still earlier. The months preceding the outbreak of the Second World War in 1939 witnessed scientific discoveries that led to the two atomic bombs on Hiroshima and Nagasaki in August 1945.[1] Two German scientists, Otto Hahn and Fritz Strassmann, first observed uranium fission in a Berlin laboratory in the winter of 1938–9, but they did not immediately understand it. Hahn wrote to Lise Meitner, his long-term colleague in Berlin but by now a refugee, and she and her nephew, Otto Frisch, also an exile, realised that what had been observed was the splitting of uranium nuclei. Otto Frisch described it as 'fission', a term borrowed from biology.

During 1939 it was found that when neutrons – uncharged particles from atomic nuclei – bombarded atoms of uranium, fission products and an immense amount of energy were released together with some spare neutrons which might then split other uranium nuclei, creating a chain reaction and releasing more and more energy. Then only two days before war broke out in September 1939, the great Danish physicist, Niels Bohr, and his American colleague, John Wheeler, published a paper explaining the fission process.

Uranium fission had previously been envisaged as a possible source of heat and power far greater than anything previously known, and as providing an explosive of unprecedented force. Bohr's paper showed that the latter was highly improbable; fission was much less likely to take place in one isotope of uranium (uranium-238) than in another (uranium-235); as natural uranium consists mainly of the former, a chain reaction will not occur in natural uranium unless the neutrons can be slowed down (or moderated) to increase their chance of hitting a fissile (uranium-235) nucleus. But slowed down neutrons will not produce the fantastically fast reactions needed for an explosion. The alternative – to separate out the 0.7 per cent of fissile uranium-235 from the 99.3 per cent of chemically identical uranium-238 – seemed impossible. A uranium bomb did not appear feasible. This was an immense relief to those who feared that some scientists remaining in Hitler's Germany after the great intellectual migration might be able to develop an atomic bomb.

Atomic power might have long-term industrial applications, but to British scientists engaged on urgent war work – such as radar research and the protection of shipping against mines and submarines – uranium research seemed unlikely to contribute to victory. However, in the spring of 1940 everything changed. A memorandum, 'On the construction of a super bomb',[2] written by two young refugee scientists at Birmingham University, Otto Frisch and Rudolf Peierls, demonstrated that 5kg of uranium-235 would suffice for the super high-speed chain reaction needed for an atomic bomb, and suggested an industrial method of separating uranium-235 from natural uranium. The atomic bomb, they wrote, would have powerful radiation, as well as explosive, effects; because radioactive substances would be widely dispersed by the wind, it could probably not be used without killing large numbers of civilians, and they thought it might therefore be 'unsuitable as a weapon for use by this country'. The British authorities promptly set up a high-powered scientific committee, code-named MAUD, which was to have an immense influence.

Soon afterwards two French scientists, Hans von Halban and Lew Kowarski, escaped to England from occupied France bringing with them, in 26 jerry cans, the world's total known stock of heavy water. They had been members of a scientific team in Paris working on slow neutron chain reactions, and the importance of their heavy water was that it was the most efficient known moderator for slowing down neutrons. Slow chain reactions in natural uranium would be useless for a bomb, but they offered the prospect of nuclear power.

Halban and Kowarski settled in Cambridge at the Cavendish Laboratory. There two of their colleagues predicted that a system using slow chain reactions in natural uranium, besides releasing energy, would create a new fissile element. It would behave like uranium-235 and would be a possible bomb material. They called it '94'; sometimes it was referred to as '239', and later it was named plutonium.[3]

The MAUD Committee presented a two-part report[4] in the

summer of 1941. It discussed the feasibility of atomic bombs using uranium-235 or possibly '94', and described and even costed the industrial plant and processes needed for separating uranium-235 and for producing '94' in a reactor. It then discussed nuclear energy as a source of power, using natural uranium as fuel and heavy water or graphite as a moderator, and it forecast that – partly because of transport savings – a 'uranium boiler' should be cheaper than a power plant fuelled by coal or oil.

The MAUD report was taken to the United States where it had a catalytic effect. The Americans at once began to take uranium research seriously, and to tackle it with vigour and a sense of urgency and purpose. In Britain a small atomic organisation code-named TUBE ALLOYS was set up. But could beleaguered Britain – under enemy air bombardment, threatened with invasion, desperately short of manpower and materials – build the big and elaborate industrial plant needed? In late 1941 the Americans proposed a joint project but the British, still ahead and over-confident, refused; they wanted information exchanges only. Given the post-war military and industrial implications they feared to put themselves into American hands by transferring the work to the United States. But by mid-1942, while the British were still struggling to build pilot units for a uranium separation plant, the Americans had made immense scientific and technological progress and the huge Manhattan Project was beginning to take shape. Now the British were keen on partnership, but the Americans were not interested and even the information exchanges ended.

At last in August 1943 Churchill persuaded Roosevelt to sign the Quebec Agreement, promising 'full and effective collaboration'.[5] It enabled British scientists to take part in the Manhattan Project, created a Combined Policy Committee, established arrangements for the joint procurement and allocation of the essential uranium, and bound the participants not to give atomic information to third parties except by mutual consent. Nearly all the British scientists working on uranium-235 or on fast neutrons for bombs joined the Manhattan Project where their contribution, though on a small scale, was a significant factor in the readiness of the two atomic bombs for use in ending the Far Eastern war. Most of the British scientists were posted to Los Alamos, the atomic bomb laboratory in New Mexico; others worked on uranium separation; however, none was allowed into Hanford, in the State of Washington, the site where the atomic reactors (or 'piles') were built to produce military plutonium.

Meanwhile the French and British scientists engaged in slow neutron work had moved from Cambridge to Canada, where a joint Anglo/French/Canadian atomic establishment was set up, directed at first by the French scientist, Halban, and then by Dr John Cockcroft (later Sir John Cockcroft, and a Nobel laureate; Cockcroft, with E. T. S. Walton, had first 'split the atom' at the Cavendish Laboratory in 1932 and had subsequently been a leading figure in radar research). After the Quebec Agreement the Americans gave the establishment in Canada their support. It developed reactors cooled and moderated by heavy water, which were valuable research tools and served also as prototypes for Canada's later CANDU power reactors.

Churchill and Roosevelt reaffirmed and extended the Quebec Agreement in September 1944 at Hyde Park, Roosevelt's country house, pledging full Anglo–American atomic co-operation for civil and military purposes after the war. But after Roosevelt's death, the Hyde Park agreement proved to be a worthless scrap of paper, and another agreement on 'full and effective collaboration' made in Washington in November 1945 between President Truman and the British Prime Minister, Attlee, turned out to be equally illusory.[6]

Most British politicians and scientists hoped for a post-war atomic partnership with the United States, but also believed that Britain must have a national atomic project.[7] In October 1945 the Labour government, under Clement Attlee, set up a research establishment to cover 'all uses of atomic energy', and recalled Cockcroft from Canada to be the director. The site chosen was the airfield at Harwell, fifteen miles from Oxford. In January 1946 another atomic establishment was announced, to undertake the production of fissile material 'for whatever purpose it might be required'. Based at Risley, near Warrington in Lancashire, it was directed by a distinguished engineer, Christopher Hinton (at 45 Cockcroft's junior by four years). He had worked at ICI before the war, and had then been deputy director general of the wartime filling factories which produced bombs and shells; he had had a hard war but there was no prospect of peacetime ease for him. Though he had no written terms of reference he was given clearly to understand that his first task was to produce plutonium for atomic weapons, as a matter of supreme national importance and urgency.

The new organisations under Cockcroft and Hinton were placed in the Ministry of Supply, a vast wartime department which provided an instant infrastructure. It did not, however, provide a firm control of the project or of the relationship between its component parts, especially the precise responsibilities of the research establishment to the production and weapon divisions. The latter was the next to be created. A more closely integrated organisation was to be created when the AEA was set up in 1954 (see Chapter 3).

Elsewhere in the Ministry of Supply, a new Chief Superintendent of Armament Research (CSAR) was appointed in 1946. William (later Lord) Penney, a young mathematics professor, had spent the war first in bomb blast and fire research during the Blitz; then in work on the artificial Mulberry harbours used in the Normandy landings; and then at Los Alamos. There his contribution was highly valued by the Americans, who tried to persuade him to stay. But he had returned to Britain for patriotic reasons and now his task as CSAR was to run the Armament Research Department and to create within it the nucleus of a top secret atomic bomb establishment. At 37, he was the youngest of the three atomic chiefs. A formal Ministerial decision to make atomic bombs was taken in January 1947 and was disclosed to Parliament in May 1948; the announcement hardly caused a ripple of reaction.

British hopes of Anglo–American co-operation faded when the Combined Policy Committee broke down in April 1946, and then they disappeared. In August 1946 the United States Congress passed the Atomic Energy Act – known as the McMahon Act – which prohibited the passing of classified atomic energy information to foreign countries (including Britain) on pain of life imprisonment or even death. The Quebec, Hyde Park and Washington Agreements of 1943, 1944 and 1945 counted for almost nothing. But the joint procurement and allocation of uranium remained in force; so did the Quebec secrecy clause. Some relaxation was achieved in 1948 by an agreement called the 'modus vivendi', but it was very limited.

Britain felt entitled to post-war atomic status and was not prepared to leave the atomic monopoly in American hands. She was determined to create a successful national project both for its own sake and as a powerful lever to gain renewed Anglo–American collaboration. The quest for interdependence became a prime objective of Labour and Conservative governments for the next twelve years, until it was at last achieved in 1958.

In 1945–46 Britain had other motives too, of power status and of defence strategy. She was recognised as the third great power, inferior only to the two superpowers, and she still had world-wide defence responsibilities and commitments. And at the end of the war, although there were American occupying forces in Germany and Austria, there was no formal United States' commitment to defend Britain or the still prostrate countries of Western Europe. They were weak and highly vulnerable, and there was no 'nuclear umbrella' until the North Atlantic Treaty Organization (NATO) was set up in 1949.

Even before the end of the war some people, including Churchill, perceived an imminent Soviet military threat. After the war ended, there seemed no immediate danger, but by 1948 the Cold War was intensifying. A Communist regime was imposed in Czechoslovakia in February; in May 1948, two-year long discussions on international control of atomic energy finally failed and the United Nations Atomic Energy Commission was wound up; in June 1948 the Soviet blockade of Berlin began, and lasted nearly a year. Then in September 1949 the first Soviet atomic bomb test was detected, to the alarm and dismay of the Americans and British (and the amazement of most of them). In June 1950 the Korean War broke out and the world situation looked increasingly dangerous. The Attlee government began a programme of rearmament and the nuclear deterrent seemed more urgently needed than ever. Pressures on Cockcroft, Hinton and Penney and their staffs – especially on the production and weapons divisions – were immense. They responded as they and others had responded in wartime; the demands of national defence would be met at all costs.

It is Hinton's organisation that especially concerns us, since it had the task of producing plutonium in atomic reactors. When he was appointed, Hinton quickly set up his headquarters at Risley, in Lancashire, and in February 1946 assembled a nucleus staff: five engineers who had worked with him in the filling factories, three of whom had been with him in ICI. None knew anything about atomic energy – nuclear engineers did not then exist in Britain – but they were experienced and dedicated men under a leader of genius. All were disciplined in the filling factory ethos; Service demands, however impossible, must be met, and atomic energy, like explosives and propellants, was dangerous, and must be handled with care.

Within six years this nucleus had been built up into a production division with some 480 scientists and engineers. Men had been trained; sites found; a complex of factories designed, constructed and put into operation. There were plants to treat uranium ores, produce uranium metal and fabricate nuclear fuel elements; piles to produce plutonium; and chemical plants to process irradiated fuel elements and extract and purify plutonium. Little more than six years after that first meeting in February 1946, plutonium for the first atomic bomb test had been delivered to Penney's staff. Work had already begun on a gaseous diffusion plant to enrich uranium. All these novel production tasks had to be carried out quickly, without an existing stock of knowledge, previous experience or access to American information. The pressure of time meant that no pilot plants could be built. The timetable was severely compressed; as building went on the construction workers were waiting for the next drawings, while the design engineers were still waiting for research data from Harwell. But all the plants were built close to time and within estimates and all, including the Windscale piles, carried out the urgent and exacting tasks required of them so that the first weapon test, code-named *Hurricane*, could take place as planned in October 1952, in the uninhabited Monte Bello Islands off the north-west coast of Australia.

2 Windscale's Origins

The first of the atomic factories[1] was the Springfields uranium plant, near Preston in Lancashire. Its tasks were to extract uranium from ore or concentrates; to purify it by precipitation, filtration and solvent extraction and reduce it to metal; finally to cast it in rods, machine it, and enclose it in aluminium cans to form fuel elements.

In a second factory, these uranium fuel elements were to be irradiated in atomic piles to produce plutonium. This factory was also to have a chemical plant where the fuel elements, after an appropriate irradiation period, would be processed to extract the plutonium from unwanted fission products and recover the slightly depleted uranium, only a small amount of which had been consumed.

Later, a third factory would be needed: an isotope separation plant to enrich uranium by increasing the proportion of fissile uranium-235 atoms. Low enrichment would make it possible to recycle depleted uranium and to produce enriched fuel elements, the latter providing any additional reactivity needed in operating a pile. High enrichment would produce uranium that could be used for nuclear weapons (or else for fast reactor fuel; in power stations of the future these fast reactors would be capable of producing more fissile material, in the form of plutonium, than they consumed).

For the second factory the main decisions required were, what kind of piles should be built and where should the factory be sited? The choice of pile and the choice of site were inextricably joined.

Although British participants in the Manhattan Project had never been admitted to the plutonium production site at Hanford, they knew that the Hanford piles were water-cooled, graphite-moderated and fuelled by natural uranium. British scientists working in the Anglo/French/Canadian atomic project in Canada, during and just after the war, designed a small reactor – GLEEP, the graphite low energy experimental pile – which was built at Harwell. It went critical in August 1947, and was followed in 1948 by the larger BEPO (British experimental pile). GLEEP and BEPO were both air-cooled, not water-cooled. But as the Americans had produced their weapons-grade plutonium in water-cooled piles, the British scientists and engineers expected to do the same. They worked on this assumption throughout 1946 but encountered severe problems of water supply, safety and siting. Where could they find sufficient cooling water? The Hanford piles had each used 30 million gallons a day, and it had to be of exceptional purity to minimise the corrosion of metal components. Then, since water absorbs neutrons strongly, any interruption in the flow would mean both overheating (due to loss of coolant) and also a sudden very large increase in the reactivity of the pile. Unless the control rods could shut it down instantly the fuel elements would vaporise and radioactivity would be widely dispersed. Extremely unlikely as these possibilities seemed, the American scientists had considered them sufficiently serious to decide that the piles – built five miles apart – must be sited 50 miles from any town of 50 000 inhabitants, 25 miles from one of 10 000 and five miles from one of 1000. Furthermore, a 30 mile four-lane highway was built, to evacuate the area rapidly in an emergency. General Groves, head of the Manhattan Project, told the British in 1946 that he 'would not be surprised to be called to the telephone any morning to hear the news that one of the piles had gone up'. No one concerned can have been under any illusion that this was to be a risk-free enterprise.

Such safety distances would eliminate most sites in Britain, except in the north and west of Scotland. Armed with reports from consultant engineers appointed to advise on sites, Hinton and his staff spent many days searching. Two locations were suggested: Harlech, on the Welsh coast, and a Scottish site between Arisaig and Morar. Hinton rejected Harlech; it would be a desecration of an historic region, he said, and in any case there were too many people living nearby. He strongly opposed any relaxation of the siting criteria. A panel set up to review the criteria concluded that production piles must be at least 40 miles from large centres of population. It studied the risks both of a major disaster and of a localised fire in a single fuel element; the latter, it said, might create a radioactive inhalation hazard for a few miles around. One Harwell physicist, W. G. Marley, even foresaw the possibility of a fire caused by the 'release of some energy stored in the graphite by virtue of the Wigner effect'.

The panel was finally left with Arisaig as the only possibility in Britain. But developing this green field site would be a formidable task. The foundation conditions were unsuitable, communications were poor, and the labour supply would be difficult. Hinton doubted if resources could be equal to the task; besides, construction would take at least four years and the search for a site had already lasted more than a year.

The problem was solved by a fresh approach to cooling. New information suggested that a gas-cooled pile could be preferable to a water-cooled one. It was known that the Americans had originally considered gas-cooled piles but, as envisaged by them, the gas would have had to be pressurised, and there would be dangers of leakages; very large fans or blowers would have been needed to circulate it; and plutonium output would have been comparatively low. So the Americans had decided in favour of water cooling.

The British had – in Canada and later at Risley – worked on gas-cooling, but they too rejected it. However, at about the end of 1946 the Risley engineers saw ingenious ways of overcoming some of the difficulties: for example, by introducing the coolant at the side of the core instead of at the bottom, and by finning the fuel cans to improve heat transfer. Gas cooling, if feasible, would eliminate difficulties of water supply and, most importantly, would ease the safety and siting problems by removing the risk of an explosive runaway reaction due to a loss of coolant. The idea was immediately attractive for producing plutonium and also promised well for electricity generation since a net recovery of useful heat would be practicable.

The Harwell scientists and engineers pursued this line of thought. Could a plutonium production pile be an *air-cooled* 'super Harwell' pile? (BEPO, built for Harwell by the production division, was not, however, a pilot plant for the production piles; it went critical after the main pile design work was complete.) With finned fuel cans a plutonium production pile could, with reasonable efficiency, be cooled by air at atmospheric pressure blown straight through the core. An air-cooled pile would be simpler to design and quicker to build than a pressurised gas-cooled one; it would avoid the corrosion problems of water cooling; it would be inherently safer and therefore easier to site. Though not suitable for generating electricity, it would be effective for producing plutonium.

Lord Portal, Controller of Production (Atomic Energy) in the

Ministry of Supply, was lost among the technical arguments, but the Hanford piles had worked and he strongly opposed any change of plan. He even suggested that the scientists and engineers who proposed a change were guilty of a dereliction of duty. But the argument in favour of Hanford-style piles was outweighed by the advantages – particularly the safety advantage – of air cooling. When Hinton told Portal that he wanted to drop the water-cooled system and concentrate on air-cooled and gas-cooled piles Portal agreed, and work on water cooling ended in April 1947. Because of the extreme urgency attached to the defence programme, air-cooled piles were to be built immediately, and the attractive pressurised gas-cooled system deferred. The latter was developed a few years later in the eight reactors at Calder Hall and Chapelcross and the nine Magnox power stations of the first civil nuclear power programme (see Chapter 3).

Initially, two air-cooled piles were authorised to supply the required amount of military plutonium; a third pile was later authorised and construction was begun, but under American pressure it was cancelled because of uranium shortages. The siting criteria, while modified for air-cooled piles, still demanded a remote site – though less remote than Arisaig – and Sellafield, an ex-Royal Ordnance Factory on the Cumbrian coast, was chosen. The new site was renamed to avoid confusion with the nuclear factory at Springfields; it was called Windscale after the bluff overlooking the river on the seaward side. But the name Sellafield always remained in use for the Windscale/Calder site locally (see Appendix VII).

DESIGN AND CONSTRUCTION

Work began there in September 1947.[2] It was a huge construction job and at its peak the workforce totalled nearly 5000 men, with over 300 architects, engineers and surveyors. At first it was hard to get labour in this remote area, but high wages and the opportunity for almost unlimited overtime soon attracted workers to 'the gold coast'. Construction work went badly for a while, but then improved so that the design work at Risley could hardly keep pace.

Design difficulties arose with some of the mechanical equipment, especially with the fuel loading machine: it never worked properly and loading could always be more quickly and efficiently done by hand. The giant blowers, unlike anything made in Britain at that time, also gave initial trouble. But the most serious and most unexpected design problems, and the worst crises, were concerned with graphite and with fuel elements, and were the result of late or inaccurate research information.

It was known that the Americans had found at Hanford that dimensional changes occurred in graphite subjected to neutron irradiation (an effect predicted by the Hungarian-born physicist, E. P. Wigner, of Princeton University). When a party of American scientists visited Britain in 1948 they gave further information about 'Wigner growth'; it occurred, they said, not along the axis of extrusion but only at right angles to it. If so, Risley calculated, their existing design for the graphite structure was impossible. The graphite blocks were already being machined, but heroic efforts in the pile design office quickly produced an ingenious new design, allowing for horizontal expansion of the blocks while holding them firm in a lattice of graphite slats.

Then in March 1949 Harwell announced new research results showing that British graphite behaved differently from American graphite and would expand along the axis of extrusion as well as at right angles to it. In time this expansion would narrow the channels through the core, making it impossible to move the fuel elements along them; the piles would have a working life of only 2½ years. The news was devastating. Another design was rapidly produced, but it would be difficult to construct and to operate, and involved adding aluminium components which would absorb extra neutrons. This problem was still being studied when, two months later, new and still tentative information came from Canada, suggesting that the expansion would be only one-fifth of that calculated from the American data, and virtually nil along the axis of extrusion. The Risley engineers had to make a prompt decision on the basis of uncertain evidence, and they decided to go ahead with the 1948 design. Later research confirmed the rightness of their decision. Wigner growth, by itself, might permit a life of 15–35 years, the scientists thought, but internal stresses might cause cracking much sooner. Hinton now believed the piles might have a life of five years, or even up to ten years.

Little thought was given to another so-called Wigner effect: an increase in potential energy due to the displacement of atoms in the lattice structure of the graphite resulting from bombardment by neutrons; it was first predicted by Leo Szilard, another Hungarian-born physicist in the United States. During a second visit by American scientists in 1948, Harwell scientists had discussed plant location hazards with Edward Teller and G. Failla, and Teller had warned that stored energy in the graphite moderator might possibly lead to a major disaster by way of a local fire involving a single fuel rod. The Harwell people, while aware, as we saw, of the danger of a local fire, thought it unlikely to lead to a major incident; they considered stored Wigner energy a small risk compared to that of loss of coolant in a water-cooled pile.

Much forethought was devoted to the problem of defective fuel cans, or cartridges. However carefully designed and tested, with 70 000 or so fuel elements in the core some would fail more easily than others. A so-called 'burst' cartridge might release highly radioactive fission products, and oxidation of exposed uranium might cause a fire; burst cartridges must therefore be quickly detected and removed. Harwell designed a burst cartridge detection apparatus: an elegant detection instrument (American in origin) consisting of a moving wire which collected the solid daughter products of gaseous fission products by electrostatic precipitation, and passed them to a Geiger counter. Risley incorporated this device in a scanning machine consisting of vertical limbs and cross-pieces – they called it a Christmas tree – which moved up and down the discharge face of the pile, sampling each channel. The Risley engineers adopted this mechanism reluctantly, however, because they rightly disliked having moving equipment in inaccessible parts of the pile.

Fear of radioactive emissions resulting from can failure led to the installation of the famous filter galleries – sometimes called 'Cockcroft's follies' – at the top of the 410ft tall Windscale stacks. Originally Harwell had advised Risley that filters would be unnecessary since the solid products of combustion would be in such small particles that they would diffuse harmlessly from a high stack. Then early in 1948 a Harwell engineer, C. A. Rennie, recorded that the Americans were going to fit air filters to the outlet ducts of their air-cooled graphite pile at Brookhaven, but there is no evidence that this information ever reached the Risley engineers. Later in 1948 Cockcroft, visiting the United States, found that the atomic site at Oak Ridge was having trouble with particulate emissions; agglomerated uranium oxide particles were falling to the ground within the site boundaries. He therefore urged that filters should be fitted to the Windscale stacks. By this time the stack of Pile No.1 had already been built to a height of 70ft, and the only possible place to install filters was now at the top; not an ideal arrangement by any means as the area of the filters would be less, and the velocity of the gas higher, than if they were at the base of the chimney. Maintenance and servicing would also be much more difficult.

After the graphite crises of 1948–49 this must have been a shattering blow to Risley, but the design office and the structural engineers overcame formidable difficulties. For each chimney, 200 tons of structural steel, as well as bricks, concrete and heavy equipment, had to be hauled up, and the filter galleries were constructed at a height of nearly 400ft. Years afterwards Cockcroft was to write, 'We found that the UO_2 particles came from the chemical stack at Oak Ridge and not from the reactor! In spite of this misinformation the filters proved to be very useful when 10 tons of uranium fuel elements were melted in the Windscale accident'.[3] (This claim is discussed later, in Chapters 7 and 9.)

The integrity and good design of the fuel elements were crucial both to the efficiency and to the safety of the piles; indeed, fuel elements were central to all pile technology. They had to be mechanically robust, and they had to be finned for efficient heat transfer, but the can must not absorb too many neutrons. (Aluminium had been chosen for the canning material for this reason.) The technical problems seemed endless: problems of casting the uranium, and of welding and sealing the can; of fin design; of chemical interaction between uranium and aluminium, solved by putting a thin graphite lining between the rod and the can; of differential thermal expansion of uranium and aluminium, solved by putting helium in the can to provide a heat transfer medium when the originally tight-fitting can expanded away from the rod; of structural strength (graphite boats were made to support the fuel elements so that they did not distort under their own weight in the channels); of testing and quality control.

The worst difficulties seemed over by the early summer of 1950. Pile No.1 was complete and the fuel elements were loaded. Then Cockcroft sent Hinton new and disturbing figures for its probable nuclear performance. The critical mass, Harwell now calculated, might be 30 per cent higher than the original estimate; at worst it might be 250 per cent higher. Thus the power rating of the pile might turn out to be less than a third of the design figure. For Pile No.2, which would have purer graphite and improved finning on the cartridges, the outlook was better, but for Pile No.1 it was bleak.

Hinton and his staff, having survived the graphite and filter crises of 1948–49, were not to be defeated now. The reactivity of the pile could be improved by reducing the amount of neutron-absorbing material in the core. This could only be done by trimming metal from the fins of the fuel cartridges. There was no time to discharge them and send them back to the cartridge shop at Springfields. They were dealt with on the spot. A team led by Tom Tuohy, the deputy works general manager at Windscale, worked at the charge hoist (a large mobile platform at the front of the pile) where, by hand, they cut a strip one-sixteenth of an inch wide from each fin. A million fins were clipped in three weeks during August and September 1950. The reactivity was also increased by reducing the cooling channel size, and the Windscale graphite shop machined 70 000 new graphite soles to be fitted to the graphite boats already made to hold the fuel elements securely in the channels.

Pile No.1 went critical in October 1950 within ten days of the target date, but its early operation fell short of its design performance by one-third. With an improved design of fuel element, Pile No.2 went critical in June 1951; it was quickly taken to 50 per cent of its design rating, and soon to 90 per cent. The two piles began their task of producing plutonium.

These piles were massive structures. Each consisted of 2000 tons of precisely machined graphite blocks in an octagonal stack, 50ft in diameter and 25ft long. The graphite which acted as the moderator also formed the pile structure. It was retained in position by spring-loaded steel members that allowed for thermal expansion. The whole core was enclosed in a 'biological shield', a box of reinforced concrete 7ft thick to give protection against the intense radiation; it was lined with steel plates to provide a thermal shield backed by lagging boxes to provide further insulation and prevent possible spalling of the inner face of the concrete.

The graphite blocks – which, as we saw, had gaps between

them to allow for dimensional changes – were pierced by horizontal fuel channels arranged in groups of four, access to each group of channels being through a charge hole, normally kept plugged, in the front shield. In the middle of each group of four was a smaller channel used for loading absorbers (to correct the power distribution) and for irradiating isotope cans for various purposes. Besides these horizontal channels there were a few vertical channels through the graphite blocks that could be used for measurements and for irradiation experiments, especially for testing materials. Each of the 3440 fuel channels contained a string of 21 fuel elements; the total charge was over 70 000 elements. They were loaded from the charge face at the front of the pile, against which the charge hoist could be moved up and down. When the cartridges were to be discharged, they were pushed through the channels into a void at the back of the pile where they fell into skips in a water duct. From there they were withdrawn to a storage pond where they were left for a period of cooling, to allow some activity to die away before the rods were decanned and fed into the chemical separation plant.

The power level of each pile was regulated by 24 horizontal rods (twelve on each side) made of boron steel; boron is a strong absorber of neutrons, and the steel contributed mechanical strength. There were twenty coarse control rods and four fine control rods, driven in and out of the pile by electric motors which moved at a slow fixed speed. They could be moved automatically in groups, or individually by a manual pushbutton control. For immediate shutdown, or in an emergency, there were sixteen vertical ('fail-safe') rods, with a neutron-absorbing capacity more than sufficient to shut down the pile; they were inserted by gravity and held in the 'out' position by electromagnets.

Cooling air was blown through the core by eight large blowers, four in each of two blower houses outside the biological shield. There were also two auxiliary (or booster) fans, and four shutdown fans; the latter were used while the pile was not running, to remove the residual heat caused by the decay of fission products. The cooling air was vented through the 410ft stacks topped by their filter galleries.

The instrumentation available to the pile operators included various instruments indicating neutron flux, blower speed and control rod position; thermocouples to measure temperatures in different parts of the pile; and numerous alarm systems and trips. To detect burst cartridges and radioactive emissions there were static air sampling devices in the outlet air-ducts, stack filter monitors, and eight mobile scanning units (the 'Christmas trees' mentioned earlier).

The scanning units (known as BCDG for Burst Cartridge Detection Gear) were on the rear face of each pile. Each unit could be set at seventeen different levels, and each had 32 nozzles able to sample the air from 32 channels at a time. The units had their own valves, precipitators and recording equipment. Their movements covered the entire face of the pile and were mechanically synchronised. The time taken for a complete scan was progressively reduced, eventually to only 57 minutes, and the pile operators were confident therefore that no failed fuel cartridge would go undetected and unlocated for long; however, the equipment would not detect a failed isotope cartridge. The static gear could detect a burst fuel cartridge very quickly but could not locate it. The stack filter monitor was an ion chamber to measure gamma activity on the stack filter and to record it in the control room; a second monitor measured activity in the stack air.

These piles – although Hinton later called them 'monuments to our initial ignorance' – were an extraordinary technological achievement for their time. His organisation, with no previous nuclear experience, had started from scratch in February 1946 and he had drawn up the overall plan for the uranium factory, piles and chemical plant by April 1947; work on the Windscale site had begun in September 1947; Pile No.1 went critical three years and one month later, almost exactly on time; Pile No.2 went critical eight months after that in June 1951. In January 1952 the first irradiated fuel rods were fed into the chemical separation plant, and on 28 March 1952, Tom Tuohy opened the reaction vessel and saw and handled the first piece of plutonium made in Britain, destined for the first British weapon test.

The Windscale piles had worked, but had not been trouble-free. After the alarms and crises of design and construction, commissioning had gone smoothly but then there had been operating problems. The first was extremely serious; it came to light during a routine shutdown of Pile No.2 in May 1952, when it was found that over 140 cartridges (at least one damaged by contact with the BCDG) had worked their way out of the core and were lodged in outlet ducts or hanging out of their channels. They had to be cleared away and a new type of graphite boat had to be speedily designed and manufactured to ensure that the cartridges stayed in place. Another worry, also in May 1952, was an abnormal and unexplained temperature rise in Pile No.2, but it was stopped by cooling air before it had spread far through the pile; a similar but more serious incident occurred in Pile No.1 in September 1952 (see Chapter 3). However, by that time Windscale had already carried out its initial programme. Plutonium for Britain's first atomic bomb test, code-named *Hurricane*, had been delivered to the weapons division in August, and the finished atomic device – not actually a bomb – was flown to Australia for the test. On 3 October, in a group of uninhabited islands off the north-west coast of Australia, it was detonated in shallow water, with a force of some 20 kilotons (equivalent to 20 000 tons of TNT).[4]

So Britain, for good or ill, became the third atomic power. As yet, however, she had no operational nuclear capacity. Even after the *Hurricane* device had been 'weaponised', warheads would not be available in significant numbers for some years and the intended delivery vehicles, the V-bombers, were not yet in production. But *Hurricane* was the beginning of Britain's independent nuclear deterrent. It was also (Churchill had said) of supreme importance because it put Britain in a far better position to secure American nuclear co-operation both in the exchange of scientific and technical information and in joint consultations on targets and methods of attack. Windscale, with its heroic efforts and brave improvisations, had played an indispensable part in this momentous undertaking.

3 After *Hurricane*

After the successful *Hurricane* test a new era in the project was opening. There were five big developments in the atomic energy project; four were in the military and civil programmes, and one (here considered first) was organisational.

A NOVEL KIND OF ORGANISATION

The organisational change[1] was largely the inspiration of Lord Cherwell. This distinguished scientist and politician had long been Churchill's scientific *éminence grise*; he had been involved in the wartime beginnings of atomic energy in Britain, and was the Minister responsible for atomic energy policy in the 1951–55 Churchill government.

The first Soviet atomic bomb test, detected in August 1949, had been a severe shock to the United States and to Britain. That Britain was more than three years behind was due, Cherwell felt sure, to Ministry of Supply management (a view which cannot be sustained now, if it ever could). He was equally convinced that it would hinder the development of civil nuclear power, and he wanted the project out of the civil service. He envisaged a radically new organisation: a corporation fully funded by the Treasury but freed from the civil service constraints that he believed had been so damaging.

Churchill was unwilling to contemplate a change before *Hurricane*, but in November 1952 he asked the Cabinet to remit the matter to a committee chaired by the Lord Privy Seal. It advised that responsibility for atomic energy should be transferred to a non-departmental body, and a second committee, under Lord Waverley (another eminent Conservative scientist–politician with much atomic energy experience) worked out the details. The Waverley report, 'The Future of the United Kingdom Atomic Energy Project,'[2] was presented to Parliament in November 1953; in June 1954 the Atomic Energy Authority Act became law, and in August the United Kingdom Atomic Energy Authority (UKAEA) came into being. It was to be run by a Chief Executive who would be Chairman of the Board (itself called the Atomic Energy Authority). Sir Edwin (now Lord) Plowden, 'the chief of all planners' – a wartime civil servant in the Ministries of Economic Warfare and of Aircraft Production, and Chief Planning Officer and Chairman of the Economic Planning Board since 1947 – was appointed as the first chairman. Sir Christopher Hinton, Sir John Cockcroft and Sir William Penney became Authority Board Members, while remaining heads of their existing organisations (renamed Industrial, Research and Weapons Group respectively). It is outside the scope of this book to deal with the setting up of the Authority, the arguments for and against, and the consequences. Here we are concerned only with the effects of the reorganisation on the Production Division/IG and its operations.

The reorganisation was a big upheaval which created a great deal of work and uncertainty for some senior staff in the project. They had to spend much time reorganising their Groups; replacing the infrastructure of common services hitherto provided by the Ministries of Supply and Works; recruiting; and discussing career intentions with staff in post, who were given two years to choose between returning to the scientific civil service or remaining in the project under new terms and conditions. It was a difficult period for the Authority management, and especially perhaps for the IG. The change coincided with a time when greatly increased military requirements and new civil nuclear demands also made an intensive recruitment programme necessary. Leonard (later Sir Leonard) Owen, next in command to Hinton, at one time dropped all other work for six months to do nothing but recruit. Hinton had reluctantly accepted the change on the assurance that it would solve his chronic staff problems by making it possible to offer competitive salaries. But as the new organisation's scales of pay were linked to civil service scales the promised relief was negligible and the Industrial and Weapons Groups continued to be short of staff, especially of engineers. Hinton continued to complain loudly, and so, more quietly, did Penney; but to no avail.[3]

THE ATOMIC BOMB PROGRAMME

In the atomic bomb programme, the next assignments for the Weapons Group[4] were to 'weaponise' the *Hurricane* device into an operational bomb; to build up a stockpile of bombs; and to design, test and manufacture improved atomic weapons to meet Service requirements while making better use of scarce plutonium. Plutonium stocks were desperately low; *Hurricane* had used up almost all the production to date. The Production Division, and Windscale in particular, were now required to produce increased quantities of plutonium. The third pile, as we saw, had been cancelled as a result of American pressure to reduce uranium demand. The two Windscale piles – which produced less plutonium than assumed in the design – would have to work flat out, but more reactors would be needed to supplement and eventually replace them. For these the scientists and engineers turned their attention to the pressurised gas-cooled graphite-moderated system that had been considered and deferred in 1947 (see Chapter 2).

Serious nuclear power studies were in hand at Harwell and Risley[5] as early as 1950, and by 1953 Harwell had produced a preliminary design for a dual-purpose reactor for plutonium and power production (christened PIPPA). The Ministry of Supply took the view that the PIPPA reactor had a vital part to play in the energy economy, and construction of the first two PIPPAs began on the Calder Hall site, adjoining Windscale, in August 1953. The PIPPAs were graphite-moderated, cooled by carbon dioxide in a pressurised circuit, and fuelled by natural uranium canned in Magnox, a magnesium alloy able to withstand higher operating temperatures than Windscale's aluminium cans; though magnesium was so pyrophoric, this alloy was chosen because of its low neutron cross-section. Then in June 1955, in response to a greatly increased demand for plutonium, a decision was taken to build six more PIPPAs, two more on the available land at Calder Hall and four at Chapelcross, across the Solway Firth in Dumfriesshire.[6]

THE CIVIL POWER PROGRAMME

Though the Calder Hall and Chapelcross reactors were built to produce plutonium for the defence programme, they also generated electricity as a by-product (and are still producing electricity in 1991). Magnox reactors could, if desired, be optimised for electricity production, and offered great attractions as civil power reactors. Anxieties about shortage of energy supplies and the state of the British coal industry were endemic in the early post-war years. There had been a fuel crisis in the bitterly cold winter of 1946/47, when people had huddled shivering round their faintly flickering gas fires and factories had to close for lack of coal. The outlook was bleak and the prospect of cheap and plentiful oil was not then in sight. In the long term coal reserves were calculated to be sufficient for 200 years, but conservation was highly desirable. At the same time coal exports were necessary for balance of payments reasons.

Hence the appeal of nuclear power was irresistible. In theory a single ton of uranium – if complete burn-up could be achieved – would release as much energy as three million tons of coal. This performance, it was hoped, might eventually be approached in fast breeder reactors. But even in the earliest thermal reactors, with relatively low burn-up, one ton of uranium was expected to be equivalent to 10 000 tons of coal. Technical and economic feasibility studies continued at Harwell and Risley, and in 1954 an interdepartmental committee under Burke (later Lord) Trend – a senior Treasury official, subsequently Secretary of the Cabinet – was set up to examine the short-term and long-term economic implications of nuclear power.

The Trend Committee concluded that the problem of coal supplies would be difficult for several years, and would remain serious for the foreseeable future. The greatest possible development of supplementary means of generating electricity – provided the cost was not prohibitive – was desirable as soon as possible. Nuclear power would not at first be economically competitive, and neither would it immediately solve the energy problem, but it would reduce the risk of fuel shortages and the dangerous dependency on coal, and would help to conserve coal reserves. It also offered prospects of valuable high technology exports. Indeed, in March 1954 a Minister had urged British firms to be prepared to take the lead in the markets of the world. 'We cannot doubt', he said, 'that just over the horizon there are immense orders waiting for all kinds of atomic plant.'[7]

The Trend report was hopeful but cautious. The government was less cautious, and approved an ambitious civil programme, announced in February 1955 in a White Paper,[8] of twelve nuclear power stations – totalling 1500–2000 megawatts – in the years 1955–65. The earlier stations were to be on the Calder

Hall (Magnox) model, but the later stations might be based on new and improved reactor types. Yet at this time there was no operational experience even of Magnox reactors, much less of other types; the first Calder Hall reactor did not come on power until August 1956, 17 months after the White Paper.

In 1957, in the aftermath of the Suez crisis, the original programme of 1500–2000 megawatts was trebled[9] to 5000–6000 megawatts, but was later rescheduled to provide 5000 megawatts capacity by 1968. The first civil programme – all Magnox – was actually completed in November 1971[10] with the commissioning of the last station at Wylfa in Anglesey; there were nine stations in all with a total net capacity of some 3730 megawatts.

At the time of the 1955 White Paper on the first civil nuclear power programme, the Government took the view that the civil stations should be built completely by private industry. In the mid-1950s, however, there was no single firm in Britain able to undertake the design and construction of a nuclear power station, and neither had the electricity authorities any expertise in commissioning and operating nuclear power plant. To create a civil nuclear power programme, a very large contribution had to be made by the AEA. With Authority assistance several industrial consortia – groups consisting basically of heavy electrical engineering, civil engineering and boiler making firms – were formed to design and build the power stations. The Authority acted as advisers and consultants both to these consortia and to the electricity authorities; the IG had to organise itself with care to separate these functions and avoid conflicts of interest. The Authority seconded experienced staff and also provided training at the Harwell Reactor School (opened in 1954) and the Calder Operations School (opened in 1957), and by attachments to Authority establishments. The electricity authorities attached fifteen or more senior staff to Windscale, where the pile group staff was responsible for training them to take charge of the civil power programme. It was the Authority, too, that procured uranium and graphite for the civil programme: Springfields was to manufacture and supply fuel elements; Windscale in due course would receive back irradiated fuel from the power stations, reprocess it and take responsibility for the resulting radioactive waste.

The Authority did not actually design, construct or operate

the power stations but, for the already heavily burdened organisation, the first civil nuclear power programme made a formidable addition to its workload, especially that of the IG; but this was not all.

FUTURE SYSTEMS

The Magnox reactors were regarded as first generation reactors, and new and more sophisticated reactor types were expected to follow soon, perhaps even in the latter stages of the 1955 White Paper programme. From the earliest days of nuclear thinking, the fast reactor – a reactor with no moderator, and using fast neutrons – had been seen as the culmination and crown of nuclear power development. Such a reactor would be very compact, would operate at high temperatures, and would be able, if required, to produce more plutonium fuel than it consumed. By using plutonium, and depleted uranium left after irradiation of uranium fuel elements in the thermal reactors, it would complete an integrated power system and would make the most economical use of the uranium (then expected to be scarce) that was so incompletely consumed in thermal reactors, with their burn-up of some 2 per cent at best. The intention to construct a small fast reactor at Dounreay in the far north of Scotland had been announced as early as March 1954. Construction of the Dounreay Fast Reactor (DFR) began in March 1955, and it was in operation by the end of 1959; not the first fast reactor in the world but the first small fast reactor designed to generate a significant amount of power.

The Authority was already exploring the possibilities of several other systems, including heavy water reactors, but the logical line of development from the Windscale piles to the Magnox reactors would lead on to improved gas-cooled graphite-moderated models, using enriched fuel and operating at higher temperatures. Planning for a prototype 'advanced gas-cooled reactor' (AGR) began in September 1957; construction began at Windscale in October 1958, the AGR was commissioned in December 1962 and it supplied electricity to the national grid in February 1963. This small AGR (producing 28 megawatts of electricity) was later the basis of the second civi' nuclear programme of 1964.[11]

These exacting development projects added yet more to the Authority's workload – again for the IG in particular – and the Group constantly diverted experienced operational staff from the production programme and its supposedly routine work. The research staff, and the tiny central safety organisation that was being built up at Risley, concentrated most or all of their attention on new and future projects, not on existing plant. The two Windscale piles were of the highest importance to the defence programme but they were seen by many as old, run-of-the-mill installations and their operation was commonly regarded as routine.

THE H-BOMB PROGRAMME

The fifth major preoccupation of the mid-1950s was the H-bomb programme.[12] Soon after the *Hurricane* test of an atomic device in October 1952 the United States detonated its first thermonuclear device at Eniwetok (known later not to be a bomb but a huge object as big as a house). The Soviet Union followed with a thermonuclear test in August 1953, and the United States then tested a very large thermonuclear bomb at Bikini on 1 March 1954. The Churchill government decided some four months later that Britain too must have thermonuclear weapons. Atomic bombs were no longer credible as the 'primary deterrent' because of the immense – virtually unlimited – destructive power of thermonuclear weapons, and the ease and speed of delivery. To 'destroy' the Soviet Union with atomic bombs would take a huge aircraft effort over several days at least, but the Soviet Union could devastate her opponent with only a few H-bombs in much less time. Britain, it was argued, must have H-bombs if her safety was not to be imperilled. The H-bomb, unlike the atomic bomb, was a 'great leveller', cancelling out the disparity between small nations and large, its possession would safeguard Britain's position in the new thermonuclear age. The H-bomb, Churchill told Parliament,[13] would revolutionise the entire foundation of human affairs. With this awesome weapon in their arsenals, he said, none of the major powers would dare resort to war; safety would be 'the sturdy child of terror' and survival 'the twin brother of annihilation'. Moreover, the possession of her own deterrent force – and this now meant H-bombs – was seen as essential to Britain's prestige and standing in the world, and especially her power to influence America's policy.

The H-bomb decision (like the civil nuclear power decision) was announced in a White Paper in February 1955.[14] Penney's staff at Aldermaston now had the urgent task of research, design and development of a thermonuclear weapon – 1958 was the original target date for testing it – as well as developing, testing, manufacturing and stockpiling various types of atomic bomb. The thermonuclear programme also made new demands on the IG staff, now called upon to produce bomb materials of which they had no experience and for which, with some help from Harwell, they had to devise new production processes from scratch. In particular they had to produce tritium (code-named AM), a super-heavy isotope of hydrogen of atomic mass 3. This was obtained by irradiating canned rods of a lithium-magnesium alloy in a reactor, and then chemically processing the irradiated rod. Later the Calder Hall and Chapelcross reactors would also be used, but in 1954 only the two Windscale piles were available to produce this crucial material. Many difficult problems of production had to be solved, but only the irradiation process is relevant here.

COMMITMENTS AND RESOURCES

Facing the future after *Hurricane*, Hinton's chief anxiety was whether the production division's staff structure was adequate both to operate the existing factories and to design, construct and operate new ones. He even asked for an inquiry under Dr Fleck of ICI to investigate the Risley position. The need to recruit the required staff – and the advantages of freeing the atomic project from the civil service and running it like a big industrial organisation – had formed a large part of the case for creating the AEA. But after the Authority was set up in August 1954 the problem remained obdurate. There was a serious national shortage of scientists and engineers. Authority pay scales were almost completely tied to those of the civil service; the Authority was not run like an industrial firm. Staffing difficulties were particularly severe in the IG (the old production division) as good engineers were scarcer than scientists; moreover, the remote location of some of the sites did not help recruitment.

The Authority considered[15] that the number of staff it took over from the Ministry of Supply in August 1954 was materially below that required for the programmes then in hand. But in the first year, in spite of strenuous recruitment efforts, there was no effective increase. Wastage was above normal, and for two years staff had the option of working for the Authority under new conditions of employment or returning to the civil service. Though most of the senior staff chose to stay, uncertainty about large numbers of middle-ranking and junior staff lasted longer, up to July 1956. At the same time, as we have seen, there were large additions in Authority commitments to the civil nuclear power programme, to new reactor systems, to greatly increased plutonium production, and to the development and production of thermonuclear weapons.

Looking back at the end of its first year, the Authority recalled that from its inception in 1945 the atomic project had been *expanding at a rate which few, if any, industrial enterprises would consider feasible*. The programmes undertaken were closely geared to a timetable; the principal threat to this timetable was lack of skilled manpower, not lack of Government funding or even material shortages (which had never been insuperable). The difficulty of recruiting and retaining sufficient skilled scientists, engineers and craftsmen had been acute for ten years and showed no signs of becoming easier. The development of atomic energy was unique in the variety of specialised skills that it demanded; the Authority (like the Ministry of Supply project before it) was in competition with industry – now with firms which were themselves trying to recruit staff in order to enter the nuclear energy field – but the Authority did not have their freedom of manoeuvre in the matter of salaries.

In the IG in particular recruitment of scientific and professional staff had done no more than keep pace with wastage, and there had been no increase in manpower to match the increased civil and military programmes. Inadequacy of staff in the Group was a constant threat to the Authority's timetables (and timetables were inexorable in the Industrial and Weapons Groups). The Authority, in its first report to Parliament in 1955, gave warning[16] that '*there is always the fear that even if the dates are achieved (as so far they have been) details will have been skimped that will be the cause of operating troubles for the future*' (my emphasis).

The defence programme after February 1955 clearly put the Weapons and Industrial Groups under heavy pressure; the civil power programme launched in the same month, while it added substantially to the Authority's workload, bore disproportionately on the IG.

Then in June 1955 the increased military demand for plutonium came as an unexpected workload. The IG met it by building six further PIPPAs, two more at Calder Hall (making four) and four at a new site at Chapelcross. The Group, if not the Authority, was now under very severe pressure.

After the setting-up of the Authority, which Hinton had been in favour of only because he was assured it would solve his recruitment and staffing problems, he repeated his anxieties forcefully again and again. But the Authority's strength – or at least the IG's strength – continued to be disproportionate to the tasks it had to undertake, and there was a growing mismatch of programmes and resources. Hinton wrote[17] after the accident:

> The difficulty is not that the staff . . . has failed to get stronger; it lies in the fact that it has not been possible to increase the strength of this staff at a rate which enables them to overtake their growing responsibilities . . . In my view the research and development programme of the Atomic Energy Authority and the programme of engineering work contingent on this is being allowed to grow far more quickly than is wise and also to grow to a total size which is not justified by the volume of business that will flow from it . . . If R&D continues to expand in the Atomic Energy Authority as it is doing at the moment the organisation is bound to blow its top off.[18]

It is revealing to compare on the one hand the commitments, and on the other the skilled manpower resources, of the production division in 1952 (a few months before *Hurricane*) and its successor, the IG, in 1957 (a few months before the Windscale accident). In 1952 there were two production reactors to run, and four sites. In 1957 the Group had three additional sites and there were ten production reactors that were in operation, being commissioned or under construction. An experimental fast reactor was being constructed at Dounreay. Old chemical plants had been extended and modified, new ones designed, built and brought into production. The civil nuclear programme had introduced large new responsibilities and had entailed a considerable diversion of effort from the IG's own plants.

This huge expansion in commitments was matched – or mismatched – by an increase of about 33 per cent in the Group's qualified scientific and engineering strength. In 1952 the old production division had had 482 professionally qualified staff at four sites.[19] But this total was for practical purposes almost doubled by the Ministry of Supply's Chemical Inspectorate (of whose staff 336 were solely employed in atomic energy work), and the Inspectorate of Electrical and Mechanical Engineering (some 80). Thus the effective strength – even without counting Ministry of Works support – was around 900. In 1957 the professionally qualified staff of the IG was about 1200, at seven sites.[20] No wonder the first Fleck report (Chapter 7) found the Group seriously under-manned.

The 'bitch goddess' success had crowned the first stage of the British atomic project, culminating in the 1952 Monte Bello test. To the decision-makers, the task and the technology had perhaps seemed deceptively easy, and great and rapid advances seemed temptingly possible. The politicians and Whitehall cannot have realised the impact of their policies on an overstretched organisation.

Questions arise which are outside the scope of this book: for example, was the *total* Authority strength adequate or inadequate for its tasks? Or was only the IG's strength inadequate? Were Authority staff deployed to best advantage? Were the commitments excessive and, if so, why? Were the priorities wrong: for instance, was the civil nuclear programme premature or over-ambitious? Was too much Authority effort devoted to future reactor systems? Was too much of its research not mission-orientated? Given the national skills shortage, would it indeed have been in the national interest, or of benefit to the economy, if the Authority had taken a bigger share of scientists and technologists than it did?

WINDSCALE AFTER *HURRICANE*

Until the Calder Hall, and later the Chapelcross, reactors were ready, the Windscale piles had to continue to provide all the plutonium for defence requirements. This was their prime task, but they had others. One was producing polonium-210 (codenamed LM) by irradiating cartridges containing bismuth oxide in the core of the pile; this radioisotope was used as a radiation source to initiate the explosive chain reaction in atomic bombs. When research and development on hydrogen bombs began, the new requirement arose for tritium (produced in the piles by irradiating rods of a lithium-magnesium alloy). Originally these were bare rods, of $\frac{1}{2}$-inch diameter, contained in a standard isotope can, but this type was very soon replaced by a larger (0.65-inch diameter) rod in an aluminium can, enclosed in a lead annulus (to add weight to the cartridge), the whole inside a second aluminium can. However, to boost tritium production – and also, apparently, because it was feared that the lead annulus might melt and jam the inner tube – in December 1956 a Mark III cartridge was designed, fabricated and loaded at short notice.[21] It consisted of a rod of alloy, 1 inch in diameter, in a single aluminium can with no outer can or lead annulus (see Appendix VI).

Besides these bomb materials, the two Windscale piles produced quantities of radioisotopes, some for use by Harwell, others – in particular, cobalt and carbon-14 – for the Authority's Radiochemical Centre at Amersham which marketed radioisotopes for medical, industrial and various research purposes. Some channels in the piles were reserved for graphite samples, so that the effects of irradiation on graphite could be monitored and investigated. Nor was this all. As time went on, the piles had increasingly to fulfil the role of materials-testing reactors and to accommodate many test rigs. They were used, for example, for testing Magnox fuel elements for the Calder Hall and Chapelcross reactors and the civil nuclear power stations of the February 1955 programme.

The Windscale staff themselves did not run the overall operational programme for the piles. That was largely in the hands of the IG's Operations Branch at Risley, which had to integrate the work of all the factories – Windscale, Springfields and Capenhurst – and to see that the various customers' requirements were met: plutonium, polonium, tritium and uranium-235 for Aldermaston; radioisotopes for Harwell and Amersham; irradiation facilities; and fuel elements for Authority and civil reactors. Bids for irradiation space in the piles were channelled through an irradiation committee at Risley, which considered all the priorities and the safety aspects, but the final decision to accept them rested, on safety grounds, with the works general manager.

To decide rightly, he and his staff needed all the relevant information, but it appeared later that they did not always have it.

Up to mid-1953 there was no difficulty in accommodating all the demands,[22] but as they increased the position began to change rapidly. Even as early as December 1951 Risley and Windscale staff were asking not how best to run the piles according to the design conditions, but how operations could be modified to meet changing demands upon them. Many steps were taken to accommodate further defence requirements in the next five or six years. All the extra items in the pile cores, especially the AM cartridges, absorbed neutrons strongly, thus reducing the reactivity of the piles (already less than the optimum because of the unexpectedly high loss of neutrons along the large fuel channels). To compensate for this reduction, and to balance the absorption of neutrons by the isotopes being irradiated in the core, a proportion of slightly enriched fuel elements was introduced in the latter part of 1953, when enriched uranium became available from Capenhurst.

The production records of the piles compared favourably with those of more conventional industrial plants. The works general manager, H. G. Davey, estimated[23] that from 1951 to 1957 they operated with only 8 per cent lost time, including scheduled shutdowns for charging and discharging fuel elements and radioisotopes and for planned maintenance. The operation of the piles was generally trouble-free but there were three problem areas: Wigner energy, burst cartridges, and the famous filters.

Wigner Energy Problems

Wigner growth (that is, the growth of graphite under irradiation) had arisen as a serious problem in the design stage, and the design of the Windscale piles had had to be radically modified to allow for it. But there was a second and related graphite phenomenon, similar to Wigner growth, that was unknown to Hinton and his staff until after the piles were in operation. So-called 'Wigner energy' is an increase in potential energy due to displacement of atoms in the lattice when they are bombarded by neutrons. Heating anneals the graphite by providing enough vibrational energy to release the atoms so that they can return to their original places in the lattice, the potential energy being released as heat. Wigner energy is stored in graphite when it is irradiated at relatively low temperatures and, unless it is released by an annealing process, it will accumulate until eventually a spontaneous and potentially dangerous release occurs, perhaps over-heating the reactor seriously.

The first intimation of trouble was on 7 May 1952[24] when an unexpected temperature rise in the upper region of the core of Pile No.2 was observed, but was not understood. It had spread only part of the way down the pile when it was checked by increasing the flow of cooling air. Then in September 1952,[25] while Pile No.1 was shut down, there was an abnormal temperature rise and smoke was seen coming from the core. It seemed that graphite or fuel elements, or both, must be burning. Restarting the blowers was the only way of cooling the pile, but the staff were afraid of causing a general conflagration in the core. However, the decision was taken to restart them and the temperature was brought under control. The smoke, it was discovered, had been due to small amounts of lubricating oil that had been carried from the blower bearings into the core.

Subsequent investigations, and some information from the United States, identified the cause of both these incidents as the spontaneous release of stored Wigner energy. If a way could not be found of dealing with the stored energy safely while continuing to operate the piles, they would have to be closed down and new piles built at great expense. Moreover there would be no plutonium for some four years, with corresponding delay to the weapons programme; even tests would be impossible.

It was decided that the small energy release in Pile No.2 in May must have affected only the top of the pile, and that a deliberate release from the bottom should be attempted. An anneal was successfully carried out in January 1953.[26] The pile was shut down; special thermocouples were installed, to sample uranium and graphite temperatures; the cooling air was shut off at 11.15 p.m. on 9 January; pile power was then raised to 4 megawatts and held there to generate nuclear heating, and the uranium and graphite temperatures slowly rose. Then two of the thermocouples showed a sudden increase and at 3 a.m. next day the pile was shut down again. By 5 p.m. that day it was judged that the energy release was completed. The shutdown fans, and soon afterwards the main blowers, were switched on to cool the graphite in preparation for restarting the pile.

After this experimental Wigner release there were periodic anneals in both the piles,[27] fifteen in all between August 1953 and July 1957, eight in Pile No.1 (see Appendix IV). Harwell scientists were present and assisted at the first two or three, but by then it appeared to them a routine matter and the Windscale staff undertook the work themselves. In late 1953 the assistant works manager at Windscale, J. L. Phillips, consulted the Thermal Reactor Design Office about the measurement of pile temperatures during Wigner releases.[28] He would like, he said, to have enough thermocouples in the graphite (perhaps several hundred) to give 'a reasonably complete overall picture of the state of the pile', and he wanted them to be readable in the control room. He also wanted to be able to measure several fuel element temperatures in the plane of maximum energy storage, as these were the limiting features of the pile in any conditions. However, the practical possibilities, dictated partly by limited access, were restricted; the best that could be done was to provide some 66 thermocouples for graphite temperature measurement during a Wigner release – there were none in place, or thought to be necessary, during normal operation – and twenty thermocouples (fewer than in normal operation) to measure uranium temperatures. Of the twenty, thirteen registered in the control room and the remaining seven on the pile roof.

The experience of Windscale staff was that each Wigner release was different. Though they were not experiments but processes essential to the continued operation of the piles, they could not be regarded in any way as routine operations, and they became progressively difficult to achieve (see Appendix IV). The fear was that if a release was not fully carried out, pockets of stored energy could be left in some regions of the pile that might repeatedly escape annealing. Such an accumulation could, it was thought, become dangerous and might lead to a spontaneous high temperature release. For this reason the pile operators were worried if the recorded graphite temperatures seemed to show that a planned release was dying away before it was complete.

By October 1955 the pile group manager considered that a satisfactory technique for Wigner energy release had been evolved.[29] The Americans had raised the temperature of the graphite in the Hanford reactors to minimise growth and so were not troubled by Wigner energy storage, and it is understandable that when Davey tried to introduce a discussion of it at an

Anglo–American meeting on reactor hazards in June 1956 they appeared not to be greatly interested.[30]

By October 1957 when the sixteenth anneal – the ninth in Pile No.1 – was undertaken, the Windscale men had much practical experience. However, their knowledge lacked a strong theoretical basis[31] since, as we shall see, research on graphite behaviour had not been given sufficient attention. More importantly perhaps, insufficient attention had been given to improving the instrumentation on the old piles. In the Authority at this time all the emphasis was on the future. For the old Windscale piles, little time and effort could be spared even by the IG's own Research and Development Branch (R & DB), let alone by Harwell, where graphite was not considered interesting. As for the small safety organisation, set up by the IG in November 1956, it was already over-loaded with work on new systems – high temperature reactors, carbon dioxide coolant and so on – and did no work on the old piles.

Burst Cartridges

The second major problem was the risk of burst fuel cartridges (see Appendix V). If a can failed and the uranium metal was exposed it would oxidise, releasing fission products into the coolant air stream, contaminating the pile and perhaps – unless removed – causing a fire in the channel: hence the vital importance of BCDG, or scanner gear. However good the design and manufacture of the fuel elements and however rigorous the quality control, with 70 000 fuel elements in each pile some can failures were inevitable, particularly as there was only a thin graphite layer to prevent the uranium from reacting with the aluminium can. Improvements were constantly sought and there were six types of cartridge in use at different times. A 'burst', it should be noted, did not imply a gross fault; it was defined as one that gave rise to an abnormally high signal on the sensitive detection gear, although on examination some of these failures were so small as to show no perceptible hole or rupture. Even so, there were only three bursts in 1951 and ten in 1952, and out of 750 000 cartridges irradiated in seven years there were less than 600 actual or suspected bursts.[32]

More troublesome than intrinsically faulty fuel elements were those that worked their way out of the core as a result of buffeting by the cooling air.³³ In May and June 1952, while Pile No.2 was shut down, over 140 displaced cartridges were found. Some were seen hanging out of the channels at the discharge face of the pile; one found in one of the skips at the back of the pile was damaged in a way that showed that it had been caught by the BCDG in coming out of the channel. Inspecting the discharge face was difficult; a periscope had to be used since the void at the back of the pile was too radioactive for anyone to enter. At first aluminium rods were threaded down the back face to stop cartridges coming out of the channels, but this was very difficult operationally. A new device was introduced, linking the fuel elements into a train that was attached to the front of each channel. This solved the problem of blowing out, but another difficulty presented itself in the summer of 1955.

Environmental surveys for radioactivity around Windscale had been carried out regularly even before the piles went into operation. Then in July and August 1955 a new survey technique, using a probe at ground level, revealed some 'hot spots' caused by large particles of uranium or uranium oxide.³⁴ Analysis showed them to be of different ages: some were about 600 days old, some 300–350 days and some 100–150 days. The source of the particles was eventually traced. Thirteen discharged fuel elements, instead of falling into the discharge duct and being removed in the skips, had overshot, fallen and lodged in the air duct at the base of the pile. In the high temperature conditions the uranium had almost entirely oxidised in the course of time. The particles should have been stopped by the pile stack filters but, when the filters were inspected, damaged units were found and it was estimated that some 50g of radioactive material could have escaped. (Later evidence suggests that this was greatly under-estimated.)

This was serious. Sir Edwin Plowden, the AEA's chairman, called a meeting to discuss it on 27 September 1955.³⁵ Senior officials were present from the Ministries concerned (Agriculture, Fisheries and Food – MAFF – and Housing and Local Government – MHLG); so were Hinton, Cockcroft and other Authority staff. Hinton explained that fuel elements could only become lodged in the air ducts when being discharged from the piles; in future such discharges would take place only once every

$2\frac{1}{2}$ years, and after each discharge the air ducts would be inspected and, if necessary, swept out. The filters had been repaired, and steps had been taken to avoid tears in future. Such incidents, therefore, would not recur.

It appeared to the Authority scientists that there was no hazard to human beings or animals but the Lord President – then the Minister responsible for atomic energy – was promptly informed and the advice of a panel of radiological consultants was sought. The panel examined the data and concluded that it was 'improbable that any particle(s) would have entered any human being. Even if it had done so the maximum dose of radiation . . . would be less than one-tenth of that which would be required to produce local damage.'[36]

Continuing surveys in October–November 1955[37] found five more particles, of recent formation, within two miles of Windscale. These were traced to five cartridge failures in Pile No.2 which had released uranium oxides. Unfortunately the BCDG, though very sensitive to gaseous fission products, could not identify uranium oxides; additional detection equipment was clearly necessary. Meanwhile remedial work had to be done on the joints between the filter frames, and an improved filter pack was urgently considered. But the problem of emission of radioactive particles in the cooling air was not solved. Every month from October 1955 to September 1957 the Windscale Technical Committee discussed the emissions, the devices introduced and the improvements in hand. Huw Howells, the health and safety manager, thought there was little chance of solving the problem by improved filtration; the trouble would have to be eliminated at source by devices to reduce the number of damaged cartridges, detect them in the void at the back of the pile, and remove them by remote handling equipment.

Then a new source of trouble was discovered.[38] Some cartridges became jammed in the scanning gear during discharge. It must be remembered that discharging 10 000 to 20 000 in a single discharge operation was a massive operational job. In January 1957 two cans were found caught in the scanning gear; the uranium was wholly oxidised, and these alone were sufficient to account for the radioactivity found in the area in the spring and summer of 1957.

During the summer the consultants recommended a large-scale biological and sampling programme, in particular for strontium-90, around Windscale. By July 1957 strontium-90 levels in milk were in some places reaching two-thirds of the permitted level for continuous consumption by infants. The Ministry of Agriculture wrote a strongly-worded official letter to the Authority,[39] recalling the history of particulate emissions from the Windscale stacks since 1955, and at the same time asked the Medical Research Council (MRC) for advice. Sir Edwin Plowden called another high level meeting[40] with the Ministries and the consultants, and Authority officials explained the remedial measures that were in hand. Professor Mayneord said that the maximum permissible body burden for human beings would be reached only after continuous exposure at these levels for many years and in his opinion there was no cause for alarm, but the situation must be carefully watched. In August the MRC advised the Lord President of the Council and the Ministry of Agriculture that a meeting of the MRC's experts had unanimously agreed that because of the level of exposure, the short period and the small population at risk, it was 'in the highest degree unlikely that any untoward effect had occurred'. Nevertheless the matter was brought to the attention of the Prime Minister, who was greatly concerned. He gave instructions that the matter should be kept secret.[41]

The Stack Filters

The filters,[42] it will be remembered, were added as afterthoughts to the Windscale stacks as a result of Cockcroft's visit to the United States in 1948. Their purpose was never envisaged as the prevention of widespread environmental contamination in case of a major reactor accident; they were simply intended to prevent particulate emissions during day-to-day operations. Why did they fail to do so? Filtering particulate material from a gas stream is a fairly common industrial problem, and one that ought to be soluble. However, the Windscale stack filters posed problems not encountered in conventional industry, quite apart from the many difficulties resulting from their position at the top of the stacks instead of at the base, where they would have been located if the decision could have been taken in time. The permissible pressure drop was strictly limited by operational requirements, since an increase in pressure drop meant a substantial decrease in plutonium production. The coolant was a huge volume, high velocity gas stream: one ton of air a second, at a speed approaching 2000ft/minute, or over 20 miles an hour. It reached the filters at a relatively high temperature, carrying dust made radioactive in the pile, as well as particles of irradiated uranium oxide of a wide range of particle size. The radioactivity of the dust made efficient filtration especially important, and also rendered the filters difficult and dangerous to handle, and to clean or dispose of after use.

On the advice of the chemical defence research establishment at Porton, the original filter pads used were of glass wool, folded into corrugations and used dry. They were intended to be washed and re-used, but were apt to tear, especially along the corrugations, and efficiency was much reduced after washing. In April 1952 a meeting[43] at Windscale reviewed the performance of the two piles, and Hinton questioned the need for the filters. All the other safety devices had worked well so far, and the filters seemed unnecessary. They were costly in pumping power and they limited plutonium output. Could they not be dispensed with? Davey, the works general manager, had misgivings and objected; Hinton deferred to his judgement and so they remained. This was fortunate; although they failed to contain all particulate emissions in normal operations, they obviously must have reduced them substantially.

Early in 1953 efforts began to design an improved filter. Paper or other combustible material was not acceptable, and glass fibres, available in various forms, were chosen. The first improved filter consisted of sheets of bonded tissue made from fine glass fibres, sprayed with mineral oil; this type had to be replaced every ten days. The pads became less efficient as time went on, especially as they became depleted in oil, so that there would be in place at any one time filter pads of varying efficiency. With the huge flow of hot air it was difficult to maintain oil on the filters. Then two much better types of filter pad were developed, which could be used for six weeks before replacement. They were made of glass fibres bonded with resin and pressed into mats $\frac{1}{2}$-inch (or later 1 inch) thick and treated with silicone oil, much more effective than mineral oil. These pads or mats were highly compressible and up to eight could be packed into a filter frame. The last type was being progressively introduced in the summer of 1957, but installing them all was bound to take some time. By the end of 1957, when the installation would be complete, Windscale hoped and believed that the environmental contamination problems of the past two years would be over. But before then the accident of October 1957 had intervened and had put Cockcroft's follies to an altogether harder test.

THE WINDSCALE MEN

How was it possible to prevent irradiated uranium from being oxidised by exposure to air for more than a few hours? By August 1957 the Windscale men had reached a point where they felt the piles could be operated safely only if a solution could be found to this problem. The minutes of the Windscale Technical Committee throughout 1957 give 'an impression of a technical struggle taking place in hopelessly adverse circumstances and against considerable odds, and lead to the conclusion that the life of these piles as production units ... was drawing to its close', although 'they might have continued to operate for some years as irradiation facilities'.[44]

These were the views, in retrospect, of Gethin Davey, a chemist who had been in charge of the Royal Ordnance Factory at nearby Drigg during the war and then had been works general manager of Windscale from its beginning. Besides Windscale, by late 1957 his responsibilities included the new sites at Calder Hall nearby and at Chapelcross 70 miles away. It would have been a heavy load for a strong man but at this time Davey was ill and in constant pain, although no one who knew him at the time considered that his efficiency was affected. He died less than three years later at the early age of 51. A greatly respected scientist and a gentle and kindly man, he was liked and admired almost universally by his colleagues and staff. In West Cumberland, where he had lived and worked since 1940, he was well known and had won the confidence and friendship of the local people. His personal reputation and standing proved their value in 1957.

His deputy and assistant works general manager was Tom Tuohy, an auburn-haired and ebullient Irishman who was to show great personal courage during the accident. He had been at Windscale since 1950, apart from two years at Springfields. It was he who had directed the emergency fin-clipping operations, and he had been responsible for commissioning the piles and the plutonium finishing plant. Under Davey and Tuohy[45] there should have been a works manager and two assistant works managers, one responsible for the pile group – the two piles and the cooling ponds – and one for the chemical plant. As discussed above, however, the whole IG suffered from serious staff shortages which had worried Hinton and Owen since 1951, and 52 of Windscale's 784 professional posts were vacant. The assistant works manager in charge of the chemical group, Tom Hughes, who had been at Windscale since 1951, was also acting works manager for the site; then, just before the accident, he was informed that he must cover a third post, that of assistant works manager for the pile group. His first visit to the pile group was when he was called to the fire on 10 October. The pile manager was Ron Gausden, who had been at Harwell for three years before moving to Windscale, where he was assistant pile manager until becoming pile manager in 1955. He was the fifth holder of that post in seven years or less; one result of the under-staffing and over-extension of the project, as we saw, was frequent changes of staff, with experienced men redeployed especially to Calder Hall, Dounreay and Chapelcross. Gausden was a young, able, middle-grade engineer, with only two professionally qualified assistants; the pile physicist who worked with him, Ian Robertson, had only two assistants. The numbers were barely sufficient to provide a 24-hour cover, without allowing for sick leave or other absences. Under these professionally qualified staff a team of technicians operated the pile controls and carried out routine duties.

Other members of the Windscale staff involved in the work of the pile group were Huw Howells, the health and safety manager, and G. D. Ireland, the chief engineer (who was responsible for a brilliant engineering exercise in which repair work to the scanner gear was carried out, under specially designed mobile shielding, in the highly radioactive void at the back of the pile). There were also the members of the Windscale R&D Branch (R & DB (W)), especially those who were concerned with graphite and with the development of the filters. There were, besides, all the other members of the Windscale workforce, too numerous to name individually, who kept this pioneering atomic factory going.

Windscale was over-worked and under-manned with inadequate research support, but it was a well-run site with hardworking and dedicated staff. They tackled their technical problems tenaciously, overcame formidable difficulties and met all the demands made on them. Shortly before the accident, when Hinton left to become the first chairman of the Central Electricity Generating Board (CEGB), he wrote a farewell message to Davey and his staff: 'Thank you so much for all that you have done since 1947. Windscale has always been my pride and joy; a really great factory, thoroughly well managed.'[46]

After Hinton went in August 1957, Leonard Owen, his longtime associate, succeeded him as managing director of IG, but the post of Member for Engineering and Production which Hinton had also filled remained vacant for six months. Instead of being at the heart of the crisis in October 1957, Hinton could only watch sadly from his new Blackfriars office and send Windscale his sympathy and good wishes. 'Although you asked me not to answer your letters', Davey wrote in reply, 'I feel I must because I would like you to know how much they meant to me in the circumstances. To have your sympathy and understanding was a tonic in itself and my face must have been most revealing when I recognised your handwriting on the envelope.'[47]

4 The Ninth Anneal

The ninth anneal of Pile No.1 was due in early October 1957. Formerly anneals had been carried out after 30 000 megawatt days of irradiation, but the Windscale Technical Committee had recently decided to extend the interval to 50 000 megawatt days. However, the pile group manager, Ron Gausden, had asked for the change to be made in two steps, to 40 000 megawatt days on this occasion and to 50 000 later. So the ninth anneal was taking place after a longer irradiation period than before. Some small pockets of graphite, mainly in the front lower part of the pile, had apparently failed to release Wigner energy in the previous anneal and in these areas the irradiation probably amounted to 80 000 megawatt days.

Gausden prepared a programme as usual – this was Pile Shutdown Programme No.79 – detailing the sequence of actions to be followed. After the Wigner release was complete and the pile had had time to cool, the opportunity would be taken of discharging and refuelling one zone of the pile. This was usual practice, to minimise the loss of production time.

Besides Ron Gausden, the people engaged in the operation were the senior pile physicist, Ian Robertson, his two assistants, Peter Jenkinson and Victor Goodwin, and three pile control engineers, with their foremen and subordinate staff. The Wigner release on Pile No.1 would not be their only preoccupation, as they were also responsible for Pile No.2 and the cooling pond.

During the Wigner release the controls would be operated by the pile control engineers and their men, acting on the instructions and under the supervision of the pile physicists, who would be on duty round the clock. There was a detailed pile operations manual but, as Wigner releases were somewhat unpredictable and no two were alike, there was no Wigner manual and no set drill; there were only brief general instructions about maximum temperatures at various stages of the anneal and actions to be taken when they were reached (see Appendix III). Within these broad guidelines the pile group manager and the physicists had to rely on their experience and knowledge of past anneals and on their professional judgement, and could consult the R & DB – where J. C. Bell was the graphite expert – if they felt they needed to. (Indeed, the pile staff must have known more about Wigner anneals than anyone else in the Authority except perhaps their predecessors in the pile group.)

On Sunday night, 6 October, all was ready for the ninth anneal to begin next morning. Less than five days later, on Friday morning (11 October), a disturbing message reached the Authority chairman, Sir Edwin Plowden, in London, and the IG's managing director, Sir Leonard Owen, who was at Dounreay. It was sent by K. B. Ross, the Group's Director of Operations, who was on a visit to Windscale.

> Windscale Pile No.1 found to be on fire in middle of lattice at 4.30 pm yesterday during Wigner release. Position been held all night but fire still fierce. Emission has not been very serious and hope continue to hold this. Are now injecting water above fire and are watching results. Do not require help at present.

Plowden immediately instructed Owen to fly to Windscale, and then wrote to the Prime Minister, who was the Minister responsible for atomic energy: 'I have to report to you a serious incident at Windscale. The facts so far are set out in the attached memorandum.' This memorandum – sent simultaneously to the Ministry of Agriculture and also issued as a press notice just after midday – stated briefly the position as it was seen at the time:

1. At 4.30 pm on 10 October it was discovered that some of the uranium cartridges in the centre of Pile No.1 at Windscale (which was at the time shut down for routine discharge of uranium and for maintenance work) had become over-heated to the point of red heat.
2. The combustion is being held. The staff are now injecting water on it from above and the temperature has started to fall.
3. Some oxidisation of uranium has occurred. The greater part of this has been retained by the filters in the Windscale chimneys. A small amount has been distributed over the Works site, and in some areas Works personnel have, as a precaution, been instructed to remain under cover. Health Physics personnel are carrying out a continuous check both on the site and in the surrounding district in order to

ensure that any increase in the amount of radioactivity would be immediately known. There is no evidence of there being any hazard to the public.
4. The type of accident which has occurred could only occur in an air-cooled open-circuit pile and could not occur at Calder Hall or any of the power stations now under construction for the electricity authorities.
5. At this stage it is not possible to give the cause of the accident. It is likely that the pile will be out of operation for a period of some months.
6. Further reports will be issued.

What had gone wrong between Monday and Thursday? What had happened to the ninth anneal: until Thursday morning, Davey said, 'one of the best Wigner releases we have had'? The next few pages trace events day by day.

MONDAY, 7 OCTOBER

Scientists from the R & DB withdrew the graphite samples they needed to examine before the anneal. One isotope channel was discharged in case temperatures were reached which might be undesirably high for this particular material; all the other channels – including those containing AM and LM cartridges – remained loaded. Gausden checked the 66 thermocouples installed (at 4ft, 6ft and 10ft back from the charge face) to measure graphite temperatures in the pile, and replaced some that were unserviceable; graphite temperatures were only measured during a Wigner release, not when the pile was running. Nineteen thermocouples to measure uranium fuel element temperatures were in their usual positions, about 16ft back from the pile face where the pile was hottest during normal operations. However, these parts of the pile were unlikely to be the hottest during an anneal because Wigner energy was at its maximum nearer the front; there might, therefore, be areas where the actual temperatures exceeded the maximum recorded temperatures. But it was not the practice to move the uranium thermocouples nearer to the charge face for a Wigner anneal, and to do so would apparently have been extremely difficult. However accurate the thermocouples themselves were – and tests had shown them to be very accurate – this was a weakness in the system.

The main blowers were switched off and at 11.45 a.m. the pile operators began running out the bottom control rods, to start the nuclear heating in the lower part of the pile. From now on the pile staff would be intent on watching the thermocouple records since the pile was, as usual, being controlled by temperature and not by power readings; for technical reasons the latter could not be relied on in the abnormal conditions of a Wigner release. These bottom rods were slowly motored out automatically at the very slow pre-set rate of 1.2c a minute until nearly 1 p.m., and were then stopped for half an hour to note the effect; withdrawal was resumed and at 2.15 p.m. the shutdown fans and their booster fans were switched off. The control rods were again stopped in order to check some graphite thermocouples which were giving odd readings, and were found to have faulty electrical connections. Movement of the control rods began once more at 5 p.m. and the pile diverged – or reached criticality – at 7.25 p.m.

The operators, under the directions of the pile physicist, now began to manipulate the control rods asymmetrically in a pattern designed to concentrate the neutron flux, and hence the nuclear heating, towards the front of the lower part of the pile where the Wigner energy storage was greatest. The aim was to reach a maximum uranium temperature of 250°C in the first instance.

TUESDAY, 8 OCTOBER

By an hour after midnight two uranium thermocouples were recording 250°C. Graphite temperatures were between 50°C and 80°C except for one thermocouple; it registered 210°C, indicating that an energy release was already taking place in that area. There now appeared to be sufficient nuclear heat in the pile to initiate a self-sustaining Wigner release throughout the graphite. It seemed to be spreading, albeit slowly, and so it was decided to run in the control rods again to shut the pile down. After a very long day, Ian Robertson, the pile physicist, went home at 2 a.m. satisfied that the anneal was going well.

Shutting down the pile was a quicker process than running it up to power, because the control rods – for good reasons – were set to run in more quickly than they could be run out. The pile was shut down by 4 a.m. The anneal should then have continued without any nuclear heating, with the release of Wigner energy providing the necessary heat to cause further releases. However, by 9 a.m. the pile staff observed most of the recorded temperatures to be either stationary or falling. It seemed to them that the Wigner release would fizzle out, leaving much of the graphite unannealed and thus liable to spontaneous releases that could cause dangerous overheating during future pile operations. They decided to run up the pile again and apply a second nuclear heating, as on at least two earlier occasions when a Wigner release was dying away. It seems that this decision had already been made by the time Robertson came back on duty soon after 9 a.m. (although it was a decision that he later himself considered correct in the circumstances). He had 'flu – there was a bad local epidemic – but stayed all day; Wednesday and Thursday he spent ill in bed. This left two assistant physicists to provide 24-hour cover.

To run up the pile, there were two different ways of moving the control rods: either automatically in groups ('group control'), or individually by push-button control. With the latter it was possible, but very difficult, to move two rods at once. The buttons had to be pressed down all the time and, said one foreman, 'it makes your finger ache to keep even one button going'. Using the buttons, the foreman who was operating the pile manoeuvred the individual rods, inching them in and out, to try to keep the temperature steady. One uranium thermocouple showed a sudden rise, within three minutes, from 330°C to 380°C; the rods were immediately run in a few centimetres and the temperature stabilised at 330–334°C. The operator's task was to keep the pile steady at 330°C, on very low power, using coarse control rods. The pile, he found, was responding very slowly indeed to the rods, and it was virtually impossible to hold it on a straight line: as he said, 'you get the thing meandering from 328°C to 336°C'. Another operator thought the pile seemed unusually 'touchy'. The pile staff were obviously very sensitive to the feel of the pile. There was, however, no suggestion that the control system was not functioning properly. Nuclear heating was continued from 11 a.m. to 7.25 p.m.

WEDNESDAY, 9 OCTOBER

The annealing process continued uneventfully throughout Wednesday morning, but by the afternoon recorded temperatures in the core were rising faster. The operating staff took the action prescribed in their instructions on what to do when the highest temperature (uranium or graphite) reached successively 360°C, 380°C, 400°C and 415°C (see Appendix III). They closed the inspection ports on top of the pile, shut the hatch at the base of the chimney, and then at 10.15 p.m. opened the four shutdown fan dampers to give an air flow through the core. This arrested the rising temperatures but, about midnight, they began to go up again.

THURSDAY, 10 OCTOBER

Soon after midnight the assistant pile physicist saw that one thermocouple (identified as No.20/53) registered nearly 400°C. The dampers were opened again for ten minutes, without much effect. At 2.15 a.m. when 20/53 reached 412°C, they were opened for a third time, for thirteen minutes; the temperature dropped, but within an hour began to rise once more. At 5.10 a.m. the dampers were opened for the fourth time, for half an hour, and the temperature fell.

Just at this point, by chance, a small increase in radioactivity was noticed on the stack instruments; something not normally observed, and not expected while the pile was shut down. Gausden was not informed and the fact was not thought worth notice because, when a closed damper was opened, or a blower started up after being switched off for a time, a small accumulation of dust was likely to be blown up the stack.

The radiological picture had been confused by the appearance on Wednesday night of unusually high radioactivity in the stack of Pile No.2. This had led Gausden to wonder if there was a burst cartridge there and if No.2 should be closed down, but there was later found to be an instrumentation fault. Meanwhile, higher-than-normal radioactivity readings on the roof of the site meteorological station were attributed to Pile No.2, not No.1. Nevertheless readings from the stack of Pile No.1 were logged hourly.

Just after midday on Thursday, the Pile No.1 dampers were opened again, for the fifth time, for fifteen minutes; yet again, for five minutes, at about 1.30 p.m. There was now a marked increase in radioactivity in the stack, which was promptly reported to Gausden. He gave instructions to open the dampers and switch on the shutdown fans to cool the pile. But it was primarily uranium and graphite temperatures that concerned the pile staff. After the shutdown fans were switched on to cool the pile, they expected graphite and uranium temperatures to converge, with the graphite temperatures falling – as they all did – and with the uranium temperatures rising at first then levelling off and falling. The highest (recorded) uranium temperature, as expected, went up quite rapidly from 340°C to 400°C, then slowly to 420°C. Then it began to rise rapidly again; something appeared to be very wrong in the pile. At 2.30 p.m. Gausden, suspecting a bad burst cartridge, ordered the main blowers to be switched on to blow the pile cold, hoping he might perhaps be able to use the BCDG to locate the trouble. But from previous experience he was not optimistic; the scanner gear was most unlikely to be operable, as it had not been designed to operate at the temperatures reached in a Wigner release. This proved to be the case again, though the gear had been moved when the maintenance engineers had taken the opportunity of the pile shutdown to work on it the previous day.

The situation looked grave. Gausden, with the pile physicists, had so far borne all the responsibility for the Wigner operation, and even at Windscale no one outside the pile group staff was aware of the turn of events. Others were soon to be involved. At 2 p.m. Gausden told Tom Hughes that Pile No.1 seemed to be in serious trouble. Hughes, because of Windscale's staff shortages, was both manager of the Windscale chemical group and acting works manager, and had just been given additional responsibility for oversight of the pile group. Already Huw Howells, the health physics manager, had been informed of a routine air sample on the site that was higher than normal; he asked for a check sample, which was also higher than normal, and he went to see Tom Hughes to see if he knew of anything that might be causing airborne contamination. Together Hughes and Howells went to the pile control room (Hughes's first visit to the pile area). At 3.45 p.m. he telephoned Davey, the works general manager, to give him the bad news about Pile No.1.

Meanwhile preparations were being made to discharge the fuel from the suspect channel but, when the plug was removed from the access hole, all four channels were seen to be red hot. More plugs were removed, revealing yet more glowing fuel elements, and they were found to be so distorted by heat that they could not be pushed out. Gausden gave orders for the extent of the fire to be determined, and then for a fire break to be created by discharging fuel channels all round the burning zone. If the fire could not be prevented from spreading throughout the core, it would have catastrophic results. A team of eight men was quickly assembled and equipped with respirators, protective clothing, film badges and dosimeters; they went urgently to work, pushing out fuel elements with heavy steel bars.

In Davey's office, people from R & DB were gathering anxiously, and K. B. Ross, the IG's Director of Operations who was at Windscale for a meeting that day, was quickly drawn into the crisis. Ross – who had been at Risley since 1954 – had limited nuclear experience but long years of experience in the oil industry, and was known as 'Abadan Ross'. He was considered a cool-headed and steady man, the sort to keep calm in an emergency. Even so, as he said a few days later, he found the picture presented by the scientists from R & DB quite terrifying. There might be a secondary, high temperature, Wigner release at 1200°C, they thought. Then the temperature of the mass of graphite might rise by 1000°C, the entire pile might be ignited, and the whole contents of the core might be released into the atmosphere and scattered over the countryside. But the highest reading passed the critical temperature of 1200°C and they all wondered what would happen if it reached 1500°C.

Davey telephoned his deputy, Tom Tuohy, at 5 p.m.: 'Come at once, Pile No.1 is on fire.' Tuohy was at home on leave, taking care of his wife and children, who all had 'flu. He went straight to the pile area. Going to the charge hoist, he saw what Gausden and others had seen; part of the pile below the channels that were being discharged was a mass of flames. On the charge hoist men were still struggling to discharge fuel elements. Back in Davey's office he discussed with Davey, Ross and the R & D scientists the various possibilities of shutting off the air, and of using carbon dioxide or argon to try to quench the fire.

About 7 p.m. Tuohy went up to the pile roof. Looking through the top inspection holes he could see a glow from the fire, down below and near the discharge face. By 8 p.m. he saw yellow flames at the back of the pile; by 11.30 p.m. there were blue flames at the back and the fire appeared to be spreading. He and Ross called on the works fire brigade to stand by with all available pumps.

The extent of the fire zone had been determined; 120 channels were involved. Around it a ring of two or three channels had been cleared as a fire-break. Now a heroic attempt was being made to eject the fuel elements from the burning channels. All the men on the charge hoist were wearing protective clothing and respirators and their faces were soon drenched in perspiration. Using every steel rod they could lay hands on, including scaffolding poles brought over from the Calder Hall construction site, the men worked tirelessly, pushing the fuel cartridges through to the back of the pile. The cartridges were so distorted that it was extremely hard to push them through, and so hot that the steel rods came out red hot and dripping with molten metal. Occasionally red hot graphite boats were pulled out; these were kicked to one side, picked up with a gloved hand and dropped safely over the side of the charge hoist into the well. Twice a cartridge was pulled out and had to be quickly pushed back into the pile. 'Nobody showed any signs of fear', the chief fire officer told me. 'You couldn't have seen a better display from the process workers. They were heroes that night.'

FRIDAY, 11 OCTOBER

About midnight Davey, ill and in pain, was persuaded to go home to bed after he, Ross and Tuohy had agreed to put water on the fire if all else failed. At 1 a.m. Ross telephoned the Chief Constable of Cumberland to tell him about the fire and warn him of a possible district emergency. The Windscale emergency procedures already agreed with the police and local authorities (Appendix VII) went into action. In the early hours of Friday the Chief Constable set up an incident centre at the works. Soon all his men were standing by at home, fully clothed and prepared for action; a fleet of buses was ready to move people away from the neighbourhood if necessary. In the Windscale factory itself, workers were warned of an emergency and were instructed to stay indoors and to wear face masks.

The fire raged on. Massive quantities of an inert gas such as argon might or might not have been effective, but only a small quantity was available. A tanker of carbon dioxide was brought from the Calder Hall site, but Tuohy was sceptical about it, since carbon dioxide had been unsuccessfully tried at Springfields to deal with magnesium fires. Between 4 and 5 a.m. on Friday the carbon dioxide was piped into the core, but to no effect. Tuohy looked through the inspection ports from time to time; the extent of the fire did not seem to be increasing, but at the back of the pile the glow was now blue rather than red, and flames darting out from the rear face were striking the steel-lined concrete wall at the far side of the air duct. Tuohy feared for the structural integrity of the biological shield of the pile on which he was standing if the fire continued much longer.

Water was a last resort. Putting water on burning graphite and metal might cause an explosive mixture of carbon monoxide and hydrogen with air, but finally – as Davey, Ross and Tuohy had agreed earlier – the risk had to be taken. Preparations were made to bring water supplies to the charge face. Equipment had to be quickly and ingeniously improvised, for there were no water connections available. There were arrangements for firefighting elsewhere, but in the pile area it had been a rule that water must be kept out of the pile building for fear of a criticality accident with the enriched uranium cartridges. Moreover, it was assumed that dowsing a channel fire with water would contaminate the pile so as to make it inoperable and destroy it as a production unit. It had been regarded as axiomatic that any fire would start in a single channel, that the radioactivity released would be quickly detected by the BCDG, and that the burning fuel element or elements would be discharged before the fire could spread. But the paradox was that the only way to cool the pile was by blowing air through it; yet if a fire had begun to take hold, the coolant air would fan the fire, not quench it.

By 7 a.m. preparations were complete. Four hoses from the Windscale fire engine were wired to scaffolding poles, ready to be pushed into the pile. They should go in, Tuohy decided, 2ft above the highest point of the fire zone. As the shifts would be changing shortly nothing could be done until the night shift had left and the day shift was under cover. Soon Davey was back on the site. Then, just before 9 a.m., the water was turned on, at first at minimum pressure. Tuohy remained in the pile building, watching and listening intently, and directing the force of the water. The critical moment passed and there was no explosion. As the water pressure increased he looked through the inspection holes and saw cascades of water pouring down, but no diminution in the mass of flames at the back of the pile.

Then, after about an hour, the shutdown fans (which had been kept in operation to maintain tolerable working conditions on the charge hoist) were switched off. The result was dramatic. Looking again through the inspection holes Tuohy saw the fire rapidly dying out. By midday he was able to report to Davey that the immediate danger was past and the situation under control. The Chief Constable of Cumberland was assured that the emergency was over and no evacuation of the neighbourhood would be necessary. Only then did Ross send his brief message to the Authority Chairman and the Group managing director.

The hoses were left playing for 30 hours, to be absolutely safe. The water flooded the forecourt, and it had to be pumped as quickly as possible into the ponds since it became very radioactive in passing through the core of the pile.

By Friday night, everyone in the London office thought the whole thing was over. Arnold Allen, the Chairman's Secretary, told me that he and his wife were at an office dance in the evening; on the way home they saw the late evening paper placards saying 'Windscale pile on fire'. 'Old stuff', they said, and went on their way.

SATURDAY, 12 OCTOBER

By Saturday afternoon the pile was cold. The Windscale men, who had shown such exemplary courage and devotion to duty all through the crisis, now faced the long labours of clearing up. They faced, too, all the other consequences of the accident, and the stress and anxiety of an official inquiry. But providentially the worst fears of Ross, Davey, Tuohy and their colleagues had not been realised, and an environmental catastrophe had been averted.

ENVIRONMENTAL CONTAMINATION

The effects of what had been a very self-contained crisis soon began to extend beyond the Windscale site boundary. The first sign of a radioactive release was at about 2 p.m. on Thursday afternoon, when dust from a routine air sampler half a mile from the pile building was found to be abnormally radioactive. Some radioactivity was also detected at the site meteorological station. The health physics staff began taking additional air samples at a dozen or so points within the factory perimeter. Howells, as we saw, realised that something was seriously wrong and he and Hughes together went to the pile control room.

It was essential to find out quickly whether there was any public health hazard beyond the site boundaries, whether from external radiation, inhalation of radioactive materials, or radioactive contamination of the food chain. Howells at once began an environmental survey. The quickest way to get information about radioactive contamination was by air-sampling and by direct measurements of external gamma radiation. Any biological monitoring programmes would have to be guided by these results. Biological monitoring would take longer because of the time involved in collecting samples and, even more, analysing them; some chemical analyses would provide fairly rapid answers – such as the concentrations of radioactive iodine and strontium – but to distinguish between the short-lived strontium-89 and the long-lived and dangerous strontium-90 would take many hours.

At 3 p.m. Howells despatched the only available health physics van along the coastal track to Seascale, apparently downwind. But where was the radioactive plume going? Light winds from the north-east appeared to be blowing it offshore and out to sea, and the Authority issued a press release to this effect. However, the weather pattern was complex and changeable, and at Windscale only the ground wind was observed, from a wind vane on the factory site.[1] Above these light, variable, easterly winds there was an inversion layer at 400ft, and above that south-west winds prevailed. Then, in the early hours of 11 October, a cold front caused the wind to freshen and veer northerly, blowing from the north-west for some twelve hours. Thus there were two distinct plumes: the earlier carrying material north-east, the later moving to the south-east over England

and eventually (in low concentration) over Western Europe.²

To return to 10 October and to the health physicists and their environmental survey; a second health physics van arrived at 5 p.m. and was sent up the coast in a northerly direction. These two monitoring vans maintained continuous patrols, for miles around the site, throughout Thursday night and the next day. Their measurements were not very accurate, but were likely to over-estimate rather than otherwise because of the effect of contamination collecting on the roofs and tyres as the vehicles travelled around. They satisfied Howells that radioactive contamination was well below district emergency levels. He decided that there was no hazard from external radiation; that intensive air-sampling showed that there was no significant inhalation hazard; and that the health risk would be solely from radioactive materials entering the human food chain. The gamma survey suggested that, if the fallout consisted of a normal mixture of fission products, some contamination of milk was likely. (Later the mixture proved not to be 'normal' since much of the volatile and short-lived iodine-131 escaped to the atmosphere, while a high proportion of other fission products, including strontium-90, was trapped inside the pile or on the filters: see Chapter 9 and Appendix IX.) Howells turned first for guidance to recent important work by the two Harwell scientists, Marley and Fry, on safety criteria for siting civil nuclear power stations, but he found it unhelpful from his point of view as it dismissed iodine-131 as unimportant.

In the course of the accident the first of the two major releases of radioactivity to the atmosphere occurred about midnight on Thursday, and shortly afterwards, when the uranium was burning. As a result of extensive air samplings throughout the Windscale and Calder site, workers had all been instructed to remain under cover; for an hour or so they had to wear face masks. The second big release was on Friday morning between 9 and 11 a.m., when the water was first put on the fire and a rush of steam carried radioactive particles and gases up the stack. Everyone on the site was ordered to remain indoors as there was no contamination inside the buildings (other than the pile area). The highest readings on Friday morning were on the Calder Hall construction site and around the chemical plants, so the construction workers were sent home and the chemical plants shut down, while workers from the plants went to sit in the canteen.

The senior medical officer, Thomas Graham (always known as 'Thos'), gave directions to the canteen staff about the safety of food.

Early on Friday morning Howells put in hand a biological monitoring programme, sampling vegetation, soil, grass and various foodstuffs. In particular he arranged to have milk samples, starting with Thursday evening milkings, collected from local farms. Obviously there would be a time-lag between the deposition of fallout on grazing land and the appearance of radioactivity in the cows' milk so, when the first results came in from the Windscale chemists on Saturday morning, it was not surprising that the Thursday evening and Friday morning milk showed only traces of iodine-131; but by mid-afternoon on Saturday Howells saw that the Friday afternoon milk showed iodine-131 levels of between 0.4 and 0.8 microcurie per litre.

He had no available standard to compare them with. The International Commission on Radiological Protection (ICRP) had made extensive recommendations about internal irradiation limits but had dealt only with limits for lifetime exposure of radiation workers: it had not yet addressed the question of radiological protection in a once-in-a-lifetime emergency for the general population (including children); and it had developed a model for 'standard man' but not for standard child or standard baby. However, Howells was aware of a recent report by Dr Scott Russell, of the Agricultural Research Council (ARC/RBC/5), recommending limits for use by the civil defence authorities in the event of a nuclear attack. The limit Scott Russell proposed for iodine-131 in milk was 0.3 microcurie per litre, calculated to keep the total radiation dose to the thyroids of infants – the most vulnerable members of the population – below 200 rads (the level at which clinical evidence of radiation damage existed). The figure was not really appropriate for the present situation but it was the only one in the scientific literature that gave Howells any help at all, and when he saw the figure of 0.4 microcurie per litre he advised the works general manager that consumption of local milk must be stopped as quickly as possible. Davey told him to get in touch with Dr Andrew McLean, the IG's chief medical officer at Risley. Now began a series of long and complicated discussions and telephone conversations that went on until after midnight.

Hitherto McLean and the small safety organisation at Risley

had been kept at arm's length. About midnight on Thursday Thos Graham, the senior medical officer at Windscale, had called McLean at home to say that there had been some kind of accident in one of the piles and radioactivity was being released, but he had no details. McLean wondered if he should go straight to Windscale; he telephoned F. R. Farmer, the chief safety officer, and together he and John Dunster, the Group's chief health physicist, went to Farmer's house. They were prepared to drive to Windscale at once; first they telephoned Windscale, but were assured by Ross that the situation was under control and help was not needed.

Early next morning, they again spoke to Ross, who again told them everything was under control. They also called Graham, who was able to give them figures for gamma radiation levels outside the chemical plant surgery, where one of the nursing sisters had been taking her own freelance measurements with a probe at the window. At noon McLean telephoned again; this time he spoke to Davey, and emphasised how concerned he, Farmer and Dunster were about the possible public health hazard. Davey, too, was reassuring; an extensive radiological survey, he said, had shown gamma radiation measurements to be less than 1 per cent of that previously agreed as requiring immediate evacuation of the area.

So far all that McLean and his colleagues had heard suggested nothing much worse than the failure of a few cartridges, but they were worried. When they saw the press statement about 1 p.m. on Friday they knew that the pile had been on fire, and they realised that it would not take many red hot cartridges to release dangerous quantities of volatile radioactive material. Anxious about the probable contamination of pasture and hence of milk, they at once arranged for Harwell to have a shift of analytical chemists on duty for the weekend, and with Howells for a fast car to take samples of Friday evening milk down from Windscale to Harwell that night.

Work on the samples began there at 8 a.m. the next day. The laboratories at Harwell were well placed to undertake this work as they were already experienced in analysing for weapon test fallout (the United States and the Soviet Union were still carrying out numerous large-scale nuclear tests in the atmosphere) and Harwell's gamma spectrometer enabled them to produce results more quickly than the Windscale chemists could.

On Saturday afternoon McLean at Risley and Howells at Windscale received the first milk results by telephone from Harwell. McLean and Dunster went to Farmer's house to discuss the implications of the 0.4 microcurie figure. They felt that 0.4 microcurie might well be above an acceptable limit for milk consumption, but they hesitated to make an immediate decision because the ICRP recommendations about iodine-131 only referred to continuous intake by radiation workers.

Meanwhile, as we saw, Howells was trying, on Davey's instructions, to contact McLean. He eventually spoke to him about 4.30 p.m. at Farmer's house, and McLean said that he would let him know what the limit should be after consulting his two colleagues. Howells for his part was certain that, whatever the limit, the iodine-131 content was now too high and an immediate milk ban was necessary. As time passed he was increasingly anxious about the practical difficulties of getting any action taken late on Saturday or on Sunday, and found the delay worrying and unnecessary. At 7.45 p.m. he again telephoned McLean and the others. McLean told him they had in mind a limit of 0.1 microcurie per litre, but they wished to consult Professor J. S. Mitchell at Cambridge and Dr Marley at Harwell, and would have a definite answer by 9 p.m. Howells thought it was less urgent to define the limit precisely than to act to stop milk that was obviously well above it. Meanwhile, McLean spoke to Arnold Allen, the chairman's secretary, to warn him of an impending milk ban. Howells explained the position to Davey and on his instructions worked out a plan of action with the Clerk to the Cumberland County Council; he then telephoned Farmer about 9 p.m.

McLean, Farmer and Dunster together had been making calculations based on a series of assumptions about the size of an infant's thyroid, the daily intake of cow's milk, the retention of iodine-131 in the child's body, the concentration of iodine-131 in the thyroid, and the total radiation dose that it would deliver. As a standard, they took a radiation dose limit of 10 rads, one-twentieth of the dose known in clinical experience to cause thyroid cancer. On this basis they calculated that, for the protection of small children, the maximum concentration in milk should be 0.1 microcurie per litre. (see Appendix VIII). Their 'standard child', McLean said later, was conceived and delivered in one evening.

About 10 p.m. on Saturday Farmer told Huw Howells he should go ahead and do whatever was necessary to ban local milk supplies, and Howells proceeded that night to make arrangements with the Whitehaven police and the manager of the local Milk Marketing Board (a fellow Welshman) to stop milk distribution from seventeen farms in the Windscale area. Meanwhile Farmer telephoned the MAFF duty officer in London, and the lines were busy up to midnight or later with complicated calls between McLean and his two colleagues, Arnold Allen, Sir Edwin Plowden, Sir Edmund Harwood of MAFF, Sir Leonard Owen, Eric Underwood (the Authority's Director of Public Relations), Marley and others. A press conference in London was arranged for the next day, Sunday, 13 October, and a special meeting of the Authority's radiological consultants was called for Tuesday, 15 October (see Chapter 5). Soon after midnight, Allen received a transatlantic call at his home from the US Atomic Energy Commission (USAEC). The post-accident phase had begun with a vengeance.

An odd feature of the events of Thursday and Friday was the remote relationship between Windscale and the senior health and safety staff at Risley. Why did Windscale not turn to them as soon as they realised that they had an accident on their hands? Why did the Windscale management – and Ross too – give them little or no information, and ignore or even deliberately exclude them until Friday afternoon?

The Windscale health and safety manager acted very much on its own initiative – he had little time to consult – and he at once co-operated when he was asked to send milk samples to Harwell, but did not turn to Risley for advice or information until Friday afternoon when Davey told him to contact McLean. Even then it was far from clear whether, for instance, Windscale could decide on a milk ban without Risley's approval.

There are several reasons for this situation. One is undoubtedly Windscale's tradition of proud independence and rugged self-help. In its brief history of tribulations and achievements Windscale – a unique and isolated site – had developed a powerful corporate loyalty and pride reminiscent of the local patriotism that Italians call *campanilismo*. Perhaps it is fanciful to think that Windscale staff looked at the tall pile stacks almost as an Italian looks at the bell tower of his parish church; but certainly the establishment's sense of identity and self-reliance was very strong, and it had a tradition of solving its own problems.

Another reason was no doubt the lack of established and well-understood Authority-wide procedures for dealing with major accidents. Emergency plans were on an individual site basis but, in 1957, of all the Authority establishments, only Windscale had an emergency plan, agreed with the local authorities, for action in the event of an 'off-site' accident (see Appendix VII). It was, of course, recognised from the beginning that Windscale was by far the greatest potential hazard, as it had thousands of times more radioactivity than any other site.

Moreover, the Windscale staff were all so heavily engaged in the crisis and its urgent practical problems that they had little or no time or attention to spare for anything else, including communications with Risley and London. This must certainly have been true of Huw Howells, with his tiny staff and his formidable responsibilities for safety both on the Windscale site and for many miles around. He was continuously busy from Thursday morning until Friday midnight, 44 hours later, and Saturday was no rest day.

Another reason that the Windscale staff did not naturally look at once to the Risley health and safety organisation was that the latter – as we shall see in Chapter 7 – was not yet well established. It was new, it was small, it was under strength and so far had been entirely occupied with new plants and projects. It had hardly begun to make its mark with operational staff in the IG, and did not yet loom large in Windscale's collective consciousness. Huw Howells, as health and safety manager, was solely responsible to his own works general manager, not to Risley.

5 Damage Assessment and Damage Control

After the fire was out and the pile cold, weeks of intense activity followed – in London and at Windscale, Risley and Harwell – to answer questions, assess and contain the environmental damage, clean up the plant, investigate the accident and its implications for the nuclear enterprise, and decide on future action. But Pile No.2 continued to run (until it was shut down on or about 17 October: see below) and within two or three days of the accident the Windscale site was operating normally except for Pile No.1.

CONSULTATIONS

London was mainly concerned with the political, health, and public relations aspects of the accident. From the time that news of the fire reached London on Friday, 11 October, Sir Edwin Plowden had kept the Prime Minister closely informed. He wrote to him twice in quick succession.[1] Macmillan replied[2] on Sunday, 13 October, 'I am most grateful to you for . . . such early and constant information about the unfortunate accident at Windscale. It was very good of you to keep me so well informed. No doubt you will discuss with me at some appropriate moment whether or not an enquiry will be required.' Plowden wrote again on Monday morning.[3] He outlined the arrangements made with the Milk Marketing Board to restrict milk supplies once local milk was found to be contaminated by radioactive iodine emitted from the pile stacks; this was short-term contamination, since iodine-131 had a half-life of only eight days, and the great bulk of radioactive emissions, he assured Macmillan, had been caught by the filters. (The performance of the filters is discussed in Chapter 9 below.) He had, he said, arranged with Sir Leonard Owen that the court of inquiry which he, as managing director of the IG, was convening, should include two – or at least one – of the outside experts whom the

Authority retained as engineering consultants. (We shall return later to the inquiry issue.)

Later that day, Monday, 14 October, Plowden held a meeting with the interested Ministries of Agriculture, Housing and Health.[4] Sir Edmund Harwood, of the Ministry of Agriculture, was anxious to be satisfied on several technical points. He also wanted a procedure to ensure quicker information on any possible restrictions on food supplies, and stressed that the legal position would need looking into, since no firm legal basis existed for the action the Authority had had to take on milk. Oddly no one mentioned, at this meeting or subsequently, that between November 1956 and January 1957 officers of the Ministry had been in touch with the IG about possible atomic site emergencies and their effect on local agriculture.[5] 'If there were to be an accident in any one of the atomic energy establishments, which resulted in the spread of any contamination whatever', they wrote, 'this Ministry would have to take steps to prevent contaminated farm produce reaching human consumption.' They enclosed a technical paper based on the recent ARC report by Scott Russell written for civil defence purposes and intended to apply to nuclear war conditions, not to a nuclear accident in peacetime.[6] On this basis the MAFF paper identified iodine-131 as the main hazard, suggested a maximum figure for iodine-131 in a child's thyroid, and calculated what degree of contamination of pasture would make the milk supply dangerous to children's health (see Appendix VIII). However the discussion between the Ministry and the Authority was not pursued, probably because the newly formed safety branch was overstretched and had more urgent priorities.

On 15 October, after milk sampling had been extended, the restricted area had to be increased from 80 to 200 square miles. People in the newly restricted areas were assured, however, that they need not worry about the milk their children had already drunk, as the level of radioactivity was much lower than it had been in the previously restricted area.

That same day, 15 October, four of the Authority's five health and safety consultants met at Risley with Authority specialist staff under McLean's chairmanship.[7] Farmer, the chief safety officer, gave them an account of the accident, the dispersal of fission products, and the contamination levels in milk from the previous Friday evening onwards. The first analyses of Friday evening samples had shown 0.4 microcurie of iodine-131 per litre of milk. McLean explained how he and his colleagues had derived the limit of 0.1 microcurie of iodine-131 per litre of milk, calculated to prevent a dose of more than 10 rads to the thyroids of young children. The consultants discussed possible methods of determining an 'action level' for restricting milk distribution. They concluded that an action level of 0.1 microcurie radio-iodine per litre of milk would correspond to a dose of approximately 7 rads to the infant thyroid, and was appropriate to the emergency (see Appendix VIII). They discussed how accurate the analyses were and how long milk restrictions would be necessary. They agreed that the banned milk could safely be used for butter and cheese but thought it would be impolitic to do so. Other foodstuffs (such as cabbage and eggs) and other radionuclides (notably caesium and barium) need not, they said, cause concern.

The *Manchester Guardian* that morning, 15 October, had published a letter[8] from a Windscale scientist, Dr Frank Leslie. He was disturbed by the lack of official warning to the public on Thursday, 10 October, about the release of radioactivity which was liable to fall on the village of Seascale, and he quoted fallout measurements that he had made there in his own garden. The incident had been quite unpredictable in its early stages, he wrote, and it would have been prudent to warn people beyond the factory boundary to stay indoors. The Authority replied officially that it had been ready to inform the public and take protective action if at any time the radiation had exceeded known safe levels; measurements were being made continuously around the district. Dr Leslie's measurements in his garden – 20 microcuries per square metre – agreed with those made by the mobile survey teams, and were less than half the radiation level accepted internationally as safe even for a continuous lifetime exposure. No.10 Downing Street asked the Authority for an explanation, and the Prime Minister reacted strongly to the Leslie letter and to a visit on 16 October by Frank Anderson, the Labour MP for Whitehaven, who drew attention to it. Mr Macmillan deplored Leslie's action, but thought it would be unwise to prevent his accepting an invitation to appear on television.

The consultants' meeting on 15 October was followed on 19

October by another meeting[9] of experts chaired by Sir Harold Himsworth, secretary of the MRC and chairman of an eminent MRC committee on radiation hazards. (This committee had published an important report in June 1956 and in 1960 was to produce another, with a substantial appendix on the Windscale accident:[10] see Chapter 9.) Two Authority consultants, Sir Ernest Rock Carling and J. F. Loutit, were present on 19 October, with five other medical experts, the head of the ARC, and W. G. Marley of Harwell (later the Authority's chief radiological safety officer). The meeting 'unanimously agreed that the emergency actions, in so far as they concerned the hazard to man, were both well-conceived and adequate and that their rapid and efficient implementation reflected great credit on all concerned'. They agreed further that the accident should be regarded as a single incident and no fallout after 12 October need be considered, and that the danger to the population from external radiation was negligible. The hazard to human beings arose solely from radioisotopes entering the food chain, but only liquid milk was of concern, and meat, eggs and vegetables could safely be ignored. It had, they said, been wise not to allow milk products (such as butter and cheese) to be made from contaminated milk. Stopping milk distribution when the concentration of iodine-131 rose to 0.1 microcurie per litre had been fully justified; however, milk should only be released for consumption when the concentration fell to 0.06 microcurie per litre and, in any similar accident in future, milk should be declared unfit for consumption immediately the iodine-131 content reached the 0.06 microcurie level.

The experts considered three other fission products: caesium-137, strontium-89 and strontium-90. Information from Windscale about caesium-137 was scanty; rapid analysis was impossible and – rightly, the MRC experts believed – it was judged unimportant in the circumstances and so had a low priority in the heavy analytical programme. They calculated that, on any reasonable assumptions, the health hazard from caesium-137 in milk would be negligible, and strontium-89, because of its short half-life (53 days), was not a matter for concern.

Strontium-90, with a half-life of 28 years, was more worrying. The strontium-90 deposited in the region came from three distinct sources: world-wide fallout from atmospheric nuclear weapon tests, the Windscale accident, and the pre-accident emissions from the pile stacks. Two months earlier, when MRC advice had been sought on the radioactive emissions from the Windscale pile stacks (see Chapter 3), a special meeting on 15 August had agreed that a concentration of 250 'sunshine units' (or SU) – that is, 250 micro-microcuries of strontium-90 per gram of calcium in milk – could be accepted provisionally as a maximum permissible level for small populations for short periods, and this advice had been given to MAFF. It was now agreed that this recommendation could be allowed to stand, but the matter must be kept under review.

Sir Harold Himsworth noted after the meeting that it was virtually certain that this concentration, even if continued, could not cause any detectable increase in bone cancer and leukaemia, but it could not be said absolutely that there was no risk. Accidents apart, he thought that unless emissions from Windscale and strontium levels in local milk could be kept down, milk production might not be advisable in a small area immediately adjacent to the piles.

However, it later became clear that what had been taken to be a long-term upward trend in strontium-90 levels in milk really reflected a short-term seasonal fluctuation. In addition it subsequently emerged that 250SU was an excessively cautious limit; by the end of 1958 further studies by another MRC committee, of dietary contamination after a nuclear accident, led to the conclusion that a limit of 2000SU would be appropriate. Its recommendations were published in the medical press in April 1959 and were reprinted in a major MRC report in December 1960 (see Chapter 9).

We shall return in Chapter 9 to the assessments of the health impact of the accident made from October 1957 onwards.

PRELIMINARY INQUIRIES

One question requiring an urgent decision in London was what board, or court, of inquiry should be instituted. Rules on the general Ministry of Supply pattern had been drawn up by Hinton in 1952, in pre-Authority days. These rules categorised accidents according to their degree of severity: whether there was loss of life, or serious damage to health or property, and whether or not radiological hazards were involved. In the case of such major accidents the inquiry was to be undertaken by a KC or judge, and a panel was briefed in advance on atomic matters, to be called on at short notice if needed. However, it proved difficult to form and impossible to maintain such a panel and the procedures were modified after much discussion in 1956.

Under the modified procedures of 1956, after an accident the head of the Group concerned[11] was to set up a board of inquiry as soon as possible; and at the same time the responsible Minister[12] must be notified. By the time the board was ready to operate, it should be known whether the Minister wished to appoint his own inquiry and, if so, the internal one could be cancelled before it had begun work.

After more discussion in January 1957 Lord Salisbury – then, as Lord President of the Council, responsible for atomic energy – signed a Direction to the Authority to report to him 'any accident involving accidental release of ionising radiation which causes (i) loss of life; (ii) serious damage to health; (iii) interference with the enjoyment of property outside the Atomic Energy Authority's premises'.[13] It will be noticed that this definition did not include *all* off-site releases of radioactivity due to an accident, and for radiological accidents the criteria of loss of life or serious damage to health were unsatisfactory because the effects might not be known until years after the event and then might be impossible to prove. Hinton himself considered the 1957 categories of accidents inappropriate.

Immediately after the Windscale accident, on the afternoon of Friday, 11 October, Arnold Allen (Plowden's Secretary) wrote[14] to No.10 Downing Street that the Prime Minister – now the Minister responsible for atomic energy – clearly need not institute an inquiry himself into an accident such as that just reported. The Authority informed the press later that day that an inquiry was to be held, but gave no details.[15] When the Prime Minister raised the matter again on Sunday,[16] the Authority's Legal Adviser was consulted. 'The decision whether to institute an inquiry of his own', he wrote, 'is a matter entirely for the Prime Minister, but I cannot think the present circumstances justify his taking such a step, and I suggest that the Chairman should advise the Minister that our own board of inquiry – particularly if it includes a consultant engineer – is adequate to ascertain the facts.'[17]

Plowden accordingly informed the Prime Minister on 14 October[18] that Sir Leonard Owen, the IG's managing director, was setting up an inquiry to include at least one outside consultant. Next day it was announced to the press[19] that the board of inquiry would consist of Sir William Penney, Member for Weapons Research and Development (as Chairman), Sir Basil Schonland, deputy director of Harwell, and two of the Authority's engineering consultants: Professor J. M. Kay, of Imperial College, and Professor Jack Diamond, of Manchester University. As Windscale was a defence plant the committee would not meet in public and neither would its report be published, but the fullest statement of the findings consistent with security would be made public. The terms of reference were succinct: 'To investigate the cause of the accident at Windscale No.1 Pile on 10 October 1957, and the measures taken to deal with it and its consequences, and to report'.

It had now ceased to be an IG inquiry, and had become an Authority inquiry. Penney was nominated chairman as one of the two scientifically qualified full-time Members of the Authority; the other was Cockcroft, but he was away in the United States. Owen was not an Authority Member but, even if he had been, it would have been inappropriate for him to chair a formal inquiry into a major accident in his own Group.

Penney was a reluctant chairman, partly because he was a nuclear weapons expert with no reactor experience, and partly because of the many urgent demands on his time.[20] The *Antler* series of weapon tests had only just been completed in Australia; a crucial thermonuclear test – *Grapple X* – was due in the Pacific in early November; and Penney was heavily engaged in separate international negotiations in Washington and Geneva. But, as Plowden told him, there was no alternative.

The Atomic Energy Executive (AEX)[21] meeting on 17 October discussed the 'mishap at Windscale', noting with appreciation the courageous and successful efforts by which the fire had been brought under control. The same day the inquiry travelled to Windscale and began its investigations; Penney hoped to have them finished in time to write the report on Tuesday, 22 October, and leave Windscale that night. He conceded that this was optimistic. 'We may be two or three days longer. It is quite clear to me that we have a really difficult job to do and we all attach such great importance to it that we are not going to rush.'[22]

There were some public anxieties about the board of inquiry: its scope might be too limited; the full facts might not come out; Authority staff might be penalised as a result of giving evidence.[23] The local MP, Frank Anderson, took his concerns to No.10 Downing Street on 17 October, then to the Authority chairman, and lastly to Hugh Gaitskell, MP, Leader of the Opposition. He was given categorical assurances by Plowden, and indeed next day a site notice was posted at Windscale saying[24] that 'any member of staff giving evidence [could] do so quite freely and frankly with a guarantee that he need fear no victimisation'. When Anderson suggested that witnesses might be accompanied by trade union or staff association representatives, Plowden agreed to consider the suggestion, but the MP subsequently heard on the BBC that trade union representation had been refused because it was a fact-finding, not a disciplinary, inquiry. He thought it 'a very great blunder'.[25]

The form of the Penney Inquiry was indeed strongly criticised by the press, especially by *The Economist* which commented on 19 October that a private Authority inquiry, however scrupulous, was not good enough; an independent judicial assessment was needed. Other papers noted that the inquiry consisted solely of Authority and ex-Authority employees.[26] After reading *The Economist's* article, Lord Hailsham, the Lord President of the Council, asked his staff: 'Does not the Statute say anything about the circumstances in which a J. E. [Judicial Enquiry] is resolved upon? Or if not could I have some indication of the scope and number of the precedents?' The Statute did not, and neither the Lord Chancellor's Department nor the Treasury Solicitor could think of any precedents (or at least any useful ones).[27] The Penney Inquiry, its report and the outcome are the subject of the next chapter.

Whatever other action was taken, the IG still had an immediate need to carry out its own investigations, and on Tuesday, 15 October, the Group management took two decisions.[28] One was to set up a technical executive committee to deal with Windscale accident problems, and the second was to initiate a special programme of research into various aspects of the accident. Hans Kronberger, the Group's director of research and development, issued urgent instructions on the experiments required. The metallurgical laboratory, at nearby Culcheth, was to study (a) the eutectics that can be formed in AM cartridges and their ignition and combustion temperatures, and (b) the combustion of graphite. Capenhurst was to set up a model graphite channel that could be heated to 400°C. Springfields was to investigate the ignition and combustion temperatures of uranium, and eutectic formation in uranium fuel cartridges. Teams at Windscale were to examine the kinetics of Wigner release in the pile structure, and gas circulation under Wigner release conditions (the latter presumably with reference to the Magnox reactors, which were cooled by carbon dioxide).

The IG's technical executive committee[29] (TX: not to be confused with the later TEC, the technical evaluation committee) was chaired by Sir Leonard Owen. It was to 'consider how our future policy is affected by the recent Windscale incident, and what action is needed on the nuclear power programme', and was to have five working parties: one (under J. C. C. Stewart) on Wigner effects; the second (under Leonard Rotherham) on the burning of uranium, magnesium, graphite and other special reactor materials; the third (under K. B. Ross) on operating techniques in Windscale and Calder reactors; the fourth (under P. T. Fletcher) on the effect on the Authority's own programme; and the fifth (under the Group medical officer, A. S. McLean) on health physics and safety procedures. It aimed to complete its deliberations before the end of 1957, and it faced the possibility of having to provide technical arguments against political suspension of the nuclear power programme.

At its first meeting on 21 October the TX received a paper about a Windscale accident technical evaluation committee to be chaired by Sir John Cockcroft, and to include outside members (Professors Diamond and Kay), Sir Leonard Owen and the chairmen of the technical working parties. The first meeting was arranged for 4 November. However, even before it met it was superseded by the TEC under Sir Alexander Fleck. But the TX and its sub-committees continued to exist, directing the various investigations whose work was later absorbed into the Fleck TEC and its working parties. Later, in January 1958, another internal committee[30] was set up by the IG to make plans for the modification and start-up of Windscale Pile No.2, taking into account the recommendations of the Fleck Committee when received.

Gradually all these activities converged, as will be seen in Chapter 7.

Meanwhile the Prime Minister and other Ministers had to be

kept constantly informed and briefed for debates in the House of Commons (on 29 October) and later in the House of Lords (21 November), and there was a spate of Parliamentary Questions to be answered from Frank Allaun, Fred Peart, George Brown, Frank Anderson and others. The Ministers of Agriculture and of Housing and Local Government were especially concerned, as was the Lord President of the Council (Lord Hailsham) who was responsible for science policy and for the MRC. He was particularly anxious to be well armed against questions when he gave a public speech in Carlisle on 26 October.[31]

Press reactions up to the end of October were mixed. Some were favourable or at least neutral, but some were critical. There was criticism (as we have seen) of the constitution of the inquiry, of unpreparedness and complacency, of under-estimation of the seriousness of the accident, and of the failure to provide a public warning during the emergency and a public information service at Windscale afterwards.

A stream of press releases had been pouring from the Authority's press office.[32] It compiled ready answers to every conceivable question about the accident, with remarkable speed considering the difficulties of collecting up-to-date technical information from Risley, Windscale and Harwell and processing it into suitable drafts. The geographical separation caused some problems.[33] While London dealt with the national press, Windscale was obviously a focus of attention, especially for the local press, local officials and organisations such as the Cumberland branch of the National Farmers' Union, and local residents. Clearly the Authority risked speaking with two voices. Rather belatedly, towards the end of October, the London office sent one of its press officers to Windscale for the duration of the emergency, to protect the over-loaded staff there and to see that no statements were made which might run counter to Authority policy. But Gethin Davey, the works general manager, was long established in the area and was greatly liked and trusted. For years it had been his policy at Windscale (for which he had sometimes been criticised) for his senior staff and himself to play a very active part in local affairs,[34] and his silence at this juncture would surely have caused surprise and disquiet. London realised belatedly the unique personal contribution he could make in maintaining confidence, and he was permitted and indeed encouraged to talk to local people and address meetings. In the immediate aftermath of the accident some members of the public and even of the Windscale workforce were certainly worried. Miners feared that radioactivity would be drawn into mine ventilation systems and cause an inhalation hazard.[35] Farmers were anxious about their crops and livestock. Boarding school proprietors were concerned about the possibility that pupils might be withdrawn. However, after Davey and other IG staff attended meetings of the National Union of Mineworkers, the National Farmers' Union, medical officers of health and general practitioners and answered questions, confidence was established. One local doctor commented that anxiety about the accident was inversely proportional to proximity to the reactor.[36]

Naturally, the farmers were apprehensive, but the accident did not affect any farm produce except milk,[37] and the Milk Marketing Board – reimbursed by the Authority – paid them at normal rates for all the milk which could not be sold, to the tune of some £60 000. In spite of forebodings, prices of cattle and sheep held up in the markets. But one or two farmers were more unfortunate; one who was anxious, for family reasons, to sell his farm and move away could not find a buyer for some time. Another farmer, Mr Wallbank, suffered a series of misfortunes to his cattle and poultry during late 1957 and 1958, which he attributed to radioactivity from the Windscale accident.[38] Less interested in compensation than in establishing the facts, he pressed his case cogently and persistently for over three years. The most puzzling injuries were pigmented lesions, some quite severe, on the muzzles of some of his cows; they had never been seen before the Windscale accident, he said, and ceased after 1958. He noted that similar lesions were observed on other farms in the neighbourhood. His MP and the President of the National Farmers' Union took up his case, and it was investigated by the MAFF laboratories, two separate groups of veterinarians and radiobiologists, and a scientific committee[39] chaired by Lord Rothschild, then chairman of the ARC. A distinguished veterinary surgeon, at the request of the National Farmers' Union, also made an independent study of all the evidence. The Rothschild committee considered the case on three occasions. It finally concluded that there was 'not a shred of evidence' to connect the problems of Mr Wallbank's farm animals with the Windscale fire or with radioactivity, and it advised MAFF in December

1958 that the matter should be closed. After more protracted correspondence, Mr Wallbank – though still not satisfied – decided in 1961 that it would be a waste of everyone's time to pursue his arguments any further.

Cumberland remained remarkably calm.[40] It would be interesting to study the many reasons for the contrasting local reactions to the 1957 Windscale fire and to the 1979 Three Mile Island accident which caused extreme local anxiety.[41] Windscale, when it was first set up, had brought employment and considerable prosperity to a severely depressed area. Cumberland had welcomed Windscale then and did not turn against it in October 1957. The establishment had brought it pride as well as economic benefit, and the grand opening of Calder Hall by the Queen in October 1956 was still a recent memory. Locally and nationally the climate of opinion was generally favourable to atomic energy; protests against nuclear weapons and weapon tests were beginning, although CND (the Campaign for Nuclear Disarmament) did not come into existence until 1958, but for the most part atomic energy was accepted and approved and great hopes were placed on its civil use.

Nevertheless, in 1958 when the future of Pile No.2 was still uncertain, the Chief Clerk to the County Council stood ready to seek an injunction against the Authority to prevent the pile being restarted[42] (an action he never needed to take: see Chapter 7).

Meanwhile in October/November 1957, at Windscale and Risley in the north, and at Harwell and the Authority's Woolwich laboratories in the south, Authority staff were coping with the practical consequences of the accident: first came the task of determining the environmental hazard and protecting the public health; no less important, the pile and the pile area had to be made safe, and the physical mess had to be cleaned up; the detailed technical investigations and experiments already referred to had to be organised and carried out as soon as possible; and, finally, the effect of the accident on the nuclear weapons programme had to be quickly assessed.

ACTION: RADIOLOGICAL SURVEYS

The threat to public health had been identified at once (as we saw in Chapter 4) as coming from iodine-131 in milk, and on

Saturday evening 12 October action had been taken to prevent the consumption of milk produced at local farms and to bring in milk supplies from outside. As Windscale extended its surveys on 13 and 14 October – taking more than 100 samples a day – it became clear that, to meet the limit of 0.1 microcurie per litre, the milk ban boundary would have to be enlarged. MAFF announced a new control area covering 200 square miles on Tuesday morning, 15 October. The risk to children who might have drunk local milk for two to three days before the ban was extended was a point to which the MRC paid special attention (see Chapter 6).

A comprehensive programme of district surveys – drawn up and supervised by a team comprising John Dunster, the chief health physicist at Risley, Huw Howells, and Bill Templeton, in charge of biological research at Windscale – began on 15 October.[43] It included air sampling up to a distance of 30 miles, and environmental surveys to measure iodine-131 using portable geological monitors (of a type used by uranium prospectors) and gamma monitors installed in cars. These surveys helped to direct the biological monitoring teams, which sampled grass, water and various foods (cabbages, Brussels sprouts, root crops, meat and eggs) as well as milk. Samples were analysed for iodine-131, strontium-89 and strontium-90, and a few for caesium-137, cerium-144, yttrium-91, ruthenium-106, zirconium-95, niobium-95, and any other radionuclides found to be significant. But the main effort was concentrated on iodine-131 in milk because of the hazard to children. Surveys carried out in the control area and its immediate surroundings were extended to the Lancashire coast, the coast of North Wales, Northern Ireland, the south of Scotland, the Isle of Man, Yorkshire and southern England. Direct readings were made on the outside of milk churns, using scintillation counters sensitive enough to measure well below the 0.1 microcurie per litre limit, and samples were collected for laboratory analysis. No results were found requiring an extension of the milk ban beyond the 200 square mile area defined on 15 October; results outside this area were extremely low, generally less than 0.002 microcurie per litre, or one-fiftieth of the approved limit.

The whole programme employed 15–20 vehicles, and by 25 October they had registered a total of 100 000 miles in ten days. Forty people were engaged in the sampling and surveys, twenty more in handling and recording the results. The analytical work stretched the Authority's radiochemical resources to the utmost and the chemists were working round the clock, some for up to 30 hours without a break. Amazingly, 300 analyses a day were being done at Windscale, and about 135 at Harwell and Woolwich. During the milk survey programme the analytical teams handled about 3000 samples. Including 100–150 chemists, and the 10–15 staff who maintained the electronic equipment, the district surveys involved some 200 people; a number which was gradually reduced as MAFF derestricted the control area in stages, on the basis of the district surveys and MRC advice. Most of the area was derestricted by 4 November and, after six weeks, on 23 November the ban was finally lifted from the immediate Windscale neighbourhood. However, monitoring of the Windscale district for strontium, by the ARC and the Authority jointly, continued for nearly three years. This monitoring programme had been set up before the accident as a result of the radiological consultants' advice in August 1957 (see Chapter 3). A comprehensive report was published in 1960.[44]

Besides the district surveys there was a programme of human measurements at Windscale and in the neighbourhood. Urine samples were collected and analysed for strontium, and adults and children were selected for thyroid measurements. Using a scintillation thyroid counter lent by Aldermaston, altogether 70 people were measured. The results are discussed in Chapter 9.

ACTION: CLEANING UP

Windscale faced a daunting task to clear up the mess.[45] Once the pile was cold, work began to clean up the pile area, to pump away the contaminated water, to make the plant safe and to plan remedial work. Some of the uranium cartridges in the undamaged part of the core were discharged but, until the skips could be removed from the water duct, it was impossible to continue discharging. On 19 October a helicopter made a survey flight across the stack, using a technique learnt by Aldermaston staff in their nuclear weapon tests in Australia. It showed that the radioactivity on the filters was decreasing rapidly and by the next day access to the filter gallery was possible. People were anxious about the risk of releasing more radioactivity, first when the filters were removed and then if the stack was left open to the air. Tentative plans were made for sealing the top of the stack with a large, specially designed, umbrella-like construction which would be lifted into place by helicopter, and meanwhile equipment was installed to spray the filters with an oil and water emulsion to prevent the release of dust into the atmosphere. In fact the umbrella gadget was never used, and after the filters had been removed the stack opening was later sealed by the simpler expedient of covering it in with very close-fitting, specially-treated timber planks.[46]

The cooling pond, to which the skips carried the discharged cartridges, was being purged, and fresh water was fed into the pond at a rate which made it possible to discharge the contaminated water into the sea. Later, when the filters were washed down, a build-up of radioactivity was expected, and water might then have to be pumped to the effluent treatment plant before it could be discharged.

Good progress was made in removing uranium cartridges from the reactor core. By the beginning of November all the cartridges had been discharged that could be removed,[47] and twenty volunteer members of the Capenhurst staff who had come to Windscale to help were engaged in decanning the rods before they were fed into the chemical separation plant. To accommodate these fuel elements, the discharge of fuel from Calder Hall reactor No.1 was postponed for two weeks.

The charge hoist on Pile No.1 was still badly contaminated, and access would be limited for months to come. As for the interior of the core,[48] both graphite and metal appeared to be undamaged within 6 ft or so of the charge face. Beyond this, damaged graphite and many channels blocked with mangled and melted metal were visible. But it was impossible to see deeper into the core, and a special television camera on a grab was needed to investigate it. Discussions about its design began with Fairey Aviation and Armstrong Whitworth.

No attempt was made to clear the blocked channels or to remove debris from the air and water ducts. Of 180 tonnes of uranium fuel in the pile, about 22 tonnes were not recovered; it was estimated that 5 tonnes had been burnt, and that 17 tonnes remained in the core. The pile was made safe by sealing the holes in the concrete biological shield and capping the stack; then monitoring equipment and alarm systems were installed, in case residual heat generated by fission products in the core should overheat the graphite and conceivably lead to a district hazard. As for Pile No.2, it had been shut down in October 1957, ostensibly to assist the Penney Inquiry[49] but also because the Authority Executive had decided that it would be unwise to continue to operate it. Hopes of restarting it were later to be disappointed, as we shall see in Chapter 7.

By 1960, after extensive decontamination of buildings – blower houses, control room, etc. – and removal of top soil from their immediate vicinity, almost the whole pile area was clean.[50] The buildings were put to good use to house laboratories, workshops, offices and stores.

THE PRODUCTION PROGRAMME: ASSESSING THE IMPACT

How did the accident affect the civil and military production programmes?[51] The former was hard hit. The Authority was already very short of general irradiation facilities and losing the capacity of even one of the Windscale piles was serious.[52] Half the annual radioisotope production for Amersham and Harwell was lost: a serious blow, especially the loss of carbon-14, which was used in medical and biological research and was one of Amersham's most important products.

The impact on the military programme was at first assessed on the assumption that Pile No.1 was definitely finished and that Pile No.2 might not be started up again. As for plutonium from Pile No.1, since the greater part of the irradiated fuel – containing approximately one year's production – had been recovered, there would be no shortfall for 1957. If it was decided not to restart Pile No.2, that fuel too could be recovered. Pipeline stocks would ensure that there would be no loss of output for at least two years. There would, it is true, be a considerable loss of production capacity for the future, but the Calder Hall reactors were being commissioned more speedily than originally programmed; indeed, if Pile No.2 was saved, plutonium output for 1957 and 1958 would exceed production forecasts. Any deficiency in the longer term could be made up from other sources, including the CEGB reactors if necessary.

More seriously, a year's production of tritium for the H-bomb programme had been lost. Assuming production was to be continued in the Calder Hall reactors, the current design of AM cartridge could not be used unless the AM cartridges in the Windscale piles could be shown to have had no part in the accident. Half the polonium-210 output for the current year had also been lost, but the loss could be made up by increasing production in Pile No.2, if it was restarted. Even if it was not, there appeared to be little difficulty in meeting Aldermaston's requirements, which were likely to be substantially reduced before long because the use of polonium-210 in initiators was to be phased out.

6 The Penney Inquiry and the First White Paper

THE PENNEY REPORT

Looking back from 1975 to the Penney Inquiry, David Peirson – Authority Secretary and secretary to the inquiry – recalled not so much the hard work as the pleasure of walking along the seashore from Seascale to Windscale on bright October mornings.[1] He and the four members of the inquiry – Penney, Schonland, Diamond and Kay – spent ten days there and accumulated a mass of information. They visited the pile area and the health physics centre; they interviewed 37 people, some repeatedly; they examined 73 technical exhibits – reports, graphs, charts, maps, drawings, log books and records of monitoring surveys – but, as the report was so urgently required, all this material could not be digested into properly edited appendices.

The atmosphere of the hearings was intense. Some witnesses must have been mentally and physically exhausted. They had just been through a traumatic experience, unsure whether an incalculable disaster could be averted. The Penney committee was rigorous in its questioning, but the chairman especially was kind and sympathetic in dealing with these shocked and weary men.

Peirson recalled the committee's discovery that if a shorthand writer was asked to do a verbatim note of a morning's proceedings, a transcript could only be expected in about a week's time. 'And what', he asked, 'was the court to do in the meantime?' By the time it reported on 26 October, it can have received very few transcripts, perhaps for the first two days' evidence, and those rather garbled. The proceedings were also recorded but there cannot have been time to play back all the hours of tape.

By dint of very hard work, the committee completed its report[2] by 26 October (see Appendix XI). It stands up well to later examination; however, not surprisingly, some points in it are questionable, whether because of misunderstanding at the time or of knowledge gained subsequently. The 31-page report was divided into eight short chapters: an introduction, events leading up to the accident, the accident itself, the measures taken to deal with it, measures taken to protect the workers and the public, conclusions, and recommendations.

The first conclusion in Chapter VII was that the primary cause of the accident had been the second nuclear heating on Tuesday, 8 October. This, the report judged, had been applied too soon and too rapidly when some graphite temperatures were seen to be falling but when a substantial number of graphite thermocouples were still showing steady increases. This second nuclear heating, it said, had led to the fire through a sequence of events. By far the most likely was that the rapid rise of fuel element temperatures had caused one or more of the fuel cans to fail; the exposed uranium had then oxidised, generating further heat; this heat then combined with rising graphite temperatures due to later Wigner releases to initiate the fire. A second possibility which the committee could not entirely reject was that the second nuclear heating had led to the failure of a lithium-magnesium cartridge, and the oxidation of the lithium-magnesium alloy could have added further heat and initiated the fire. Once a cartridge had failed – whether uranium or lithium-magnesium – the burning of graphite would add to the heat released and contribute to the development of the fire. The evidence, the report concluded, indicated that the fire had been initiated by the failure of a fuel element can in a region just below the middle, and towards the front, of the pile.

The report's second conclusion was that the steps taken to deal with the accident, once discovered, were 'prompt and efficient and displayed considerable devotion to duty on the part of all concerned'. While the committee had naturally concentrated on what had gone wrong, it paid tribute to the efficient and energetic way in which the accident had been dealt with, and to the efforts of the Windscale staff by which a worse accident had been averted.

The third conclusion was that measures taken to deal with the consequences of the accident were adequate, that there had been 'no immediate damage to the health of any of the public or of the workers at Windscale', and that it was 'most unlikely that any harmful effects [would] develop'. However, the several responsibilities of the Group medical officer, the safety officer and the health physics manager at Windscale were not clearly defined, and tolerance levels which should have been predetermined had had to be hastily worked out after the accident.

The committee made some severe comments on the technical and organisational inadequacies it had observed, particularly the instrumentation deficiencies, the lack of an operating manual for Wigner anneals, and the poorly defined division of responsibility both within the IG – between the Operations Branch, the R & DB and other technical branches – and between the Group and Harwell. Technical changes made by one team within the Authority had not always been known to others who should have been aware of them for the proper discharge of their duties. The operations staff at Windscale were not well supported in all respects by technical advice. Changes in operating procedures, generally tending to push pile temperatures upwards, had been made without complete realisation of all the technical factors involved. After the Windscale piles had been handed over from the design offices to the operations staff, other onerous demands on the IG had meant that insufficient technical attention was available to ensure safe operation of the piles. The Windscale organisation was not strong enough to carry the heavy responsibilities laid upon it; the works general manager's responsibilities included the eight reactors built or planned at Calder Hall and Chapelcross as well as the whole Windscale site, and no one man could be expected to exercise day-to-day operational control of such a vast organisation, especially as several senior posts under him in the Windscale complement were unfilled at the time of the accident.

Finally the Penney report identified some important matters that it had not dealt with. 'Conscious of the great public anxiety concerning this accident', it said, 'we felt we should report as soon as we had been able to consider the technical evidence sufficiently to discharge our terms of reference. We have not, however, in ten days been able to make a full technical assessment ... of this matter. Moreover we are not properly constituted to recommend detailed organisational changes.' It therefore recommended, first, a technical evaluation working party within the Authority, to study urgently and thoroughly all the technical information to be derived from the accident. Second, the Authority should review the organisation and staffing of the IG in relation to its responsibilities. Third, responsibilities within the Authority for the control of health and safety should be clarified. Fourth, action should be taken to ensure that maximum permissible levels were laid down for all radioactive substances, for short term as well as for continuous exposures (a matter for which the Authority was not solely or even primarily responsible). Lastly, Pile No.2 should not be restarted until its instrumentation had been made fully adequate for Wigner anneals and until all the factors involved in the controlled release of Wigner energy had been carefully reviewed.

On 28 October the AEX[3] considered the Penney report. It decided to set up a technical evaluation committee (to be chaired by Sir John Cockcroft) and two other committees to carry out the two organisational studies recommended by the Penney report. It advised the AEA Board to accept the report and to recommend to the Prime Minister – subject to Ministry of Defence views – that it be published in full and that the MRC be invited to comment on the two chapters on health and safety of workers and the public. A submission to the Prime Minister was to be drafted for the Board's approval.[4]

Later that day the Authority sent the Penney report to Downing Street with a covering letter[5] saying that Sir Edwin Plowden thought it important that if at all possible the Prime Minister should read it before their meeting next morning. His secretary put the report before Mr Macmillan with a note suggesting that he need not read it all, as it was very technical. But by next morning he had made many detailed annotations in red pencil: 'I have read all this, it is fascinating. The problem is two-fold. (a) What do we do? Not very difficult. (b) What do we *say*? *Not* easy.'[6] He wrote at once[7] to the Lord President of the Council, Lord Hailsham, about the report, asking him to obtain the MRC's views on the health and safety aspects: a request which Hailsham passed on to Sir Harold Himsworth as a matter of great urgency.[8] The same day, 29 October, the report was circulated to Authority Members and copies were despatched to numerous Ministers and senior officials (and to Sir Alexander Fleck, Chairman of ICI) with a covering letter[9] saying that the Chairman of the Authority had it in mind to publish the report in full, but meanwhile it should be treated as confidential.

In the House of Commons, the Prime Minister was urged both to publish the full Penney report and to set up another, independent, inquiry to investigate all the circumstances. Pressed by Hugh Gaitskell, Leader of the Opposition, and other MPs –

The Penney Inquiry and the First White Paper

George Brown, Fred Peart and Frank Anderson – to publish, the Prime Minister asked for a little time to assess the Penney report and told the House about the urgent health study to be undertaken by the MRC.[10]

The Ministry of Defence, which had already been consulted about publication, gave its approval on 29 October, in a letter from the Chief Scientific Adviser, Sir Frederick Brundrett, to Sir William Penney.[11] He wrote:

> My dear Billy, I return herewith the copy of your report on the Windscale accident which you left with me yesterday. I have arranged for a letter to go – indeed it has already gone – to Edwin Plowden saying that there is no security objection to the publication of this report in its present form. Do allow me to congratulate you on the report. It seems to me to be a very fine effort. Yours ever, Freddie.

On 30 October, the Authority Board held a special meeting[12] to discuss the report, the AEX recommendations and a submission[13] to the Prime Minister drafted by the Authority's economic adviser and head of the Programmes Branch, J. A. Jukes. This draft gave two main reasons for the accident: weaknesses in the Authority's organisation and weaknesses in the instrumentation provided at Windscale so that the operating staff had insufficient information for the operation they were performing. The accident had been due to avoidable factors. The fault was not that of any one individual but was collective; no disciplinary action was being taken, and the Authority accepted full responsibility. The only significant consequence outside the Authority had been the need to destroy milk affected by the release of radioactive iodine. The draft submission went on to say that a committee – to be chaired by Sir Alexander Fleck – was being formed to make the two organisational studies proposed by the Penney report, while a technical evaluation committee, including outside experts, was to be chaired by Sir John Cockcroft. Finally, the draft explained that the accident in Windscale Pile No.1 could not possibly occur in a Magnox reactor, and there was no cause for alarm over the civil nuclear power programme or the Authority's plutonium production reactors at Calder Hall and Chapelcross.

It could be inferred from the report, the AEA concluded, that the accident might well have been very much worse, that a similar or even worse accident might have occurred on several occasions in the previous few years, and that this state of affairs, and the accident, could be directly attributed to serious defects in the Authority's organisation and to equally avoidable defects in the instrumentation of the Windscale piles. The Board took a frank and courageous stance. It recognised that the report, if published, would severely shake public confidence in the Authority's competence to undertake the tasks entrusted to it and would provide ammunition for people who had doubts about the development and future of nuclear power. There were obvious implications for the civil nuclear power programme; but, as far as the Authority was concerned, there must be no apparent attempt to gloss over the events. All that was to be stated must be stated at the outset; there must be no successive announcements each amplifying the one before. Every opportunity should be taken to explain to the public, in simple language, the meaning of Wigner energy, the design improvements in the Calder Hall reactors and the more advanced reactors to be built for the Central Electricity Authority (CEA), and the health and safety aspects of the CEA power stations. The Board asked Cockcroft, Owen and Jukes to rewrite a simpler and more positive explanation of the differences between the Windscale piles and the CEA reactors, and invited the Chairman to submit the report formally to the Prime Minister with its recommendations.

That same day, 30 October, after a discussion[14] with Sir Norman Brook (the Cabinet Secretary) and Sir Edwin Plowden, Mr Macmillan had decided most emphatically that the Penney report was not to be published. He would make a short oral statement to the House of Commons during the following week and would at the same time make available a longer written statement, probably in the form of a White Paper. It would explain that, as the Penney report was one 'made by a servant of the Authority to the Authority about a mishap at a factory engaged in defence work', it would not be suitable for it to be published. The White Paper would describe the mishap in non-technical terms, stressing the impossibility of such an occurrence in any of the civil power stations. It would say that the Prime Minister, at the suggestion of the Authority, had asked Sir Alexander Fleck to undertake a further inquiry covering the technical, health and safety, and organisational aspects of the accident, and three committees would assist him. The White Paper would probably also contain the MRC's views.

The Downing Street meeting on 30 October agreed that it was extremely important that there should be no leakage of the Penney report. On the Prime Minister's instructions all but two or three copies of the report were called in[15] from Ministers and officials, and HMSO – which already had the text in anticipation of approval to publish – was ordered to surrender the printer's copy and proofs and to break up the type at once. Even Lord Citrine (chairman of the CEA), who had the greatest possible interest in anything affecting the safety of the civil nuclear power programme which had been imposed on the CEA in 1955, was not permitted to have a copy;[16] neither was Sir Christopher Hinton, his successor-designate (as first chairman of the CEGB) and, until a few weeks before, Authority Member for Production and Engineering and head of the IG.[17]

The Prime Minister's opinion was strongly supported by the Minister of Power, Lord Mills, who wrote to him on 1 November:[18]

> I read this report before Sir Edwin Plowden came to see me today. I had formed the view, which I indicated to Sir Edwin, that it would be undesirable for the document in its present form to be published. Much of the matter seems to me appropriate only for internal consumption. It should not be impossible to have a statement indicating in general terms the cause of the accident and incorporating the reassuring paragraphs on personal and public well being.

Macmillan wrote in his diary on 30 October:

> The problem remains, how are we to deal with Sir W Penney's report? It has, of course, been prepared with scrupulous honesty and even ruthlessness. It is just such a report as the Board of a Company might expect to get. But to publish to the world (especially to the Americans) is another thing. The publication of the report, as it stands, might put in jeopardy our chance of getting Congress to agree to the President's proposal.[19]

Why had the Prime Minister taken this decision, despite the opinion of the Ministry of Defence and the Authority's courageous recommendation to 'publish and be damned'? The Authority learnt his reasons a few days later, when the Board met on 4 November.[20] Plowden told them that the Prime Minister believed that it would be wrong to publish so much detailed technical information about a defence installation. (This had indeed been said when the Penney Inquiry was announced two weeks earlier.) Even if there was no security objection to publishing so much technical detail, Mr Macmillan thought there was still a danger that it would be quoted out of context and misused in other ways by hostile critics. In particular it would provide ammunition for American opponents of the proposed amendments to the McMahon Act which would make possible the desired Anglo–American collaboration in military applications of atomic energy. (The atomic attaché at the Washington Embassy had very recently warned Plowden's office that the amendment of the McMahon Act faced 'rough going' in Congress,[21] partly because of a television programme in which a senator had – wrongly – accused Britain of refusing information to the United States about the Windscale accident.) Moreover, publishing the Penney report, the Prime Minister thought, might well adversely affect Anglo–American co-operation in other, non-nuclear defence fields.

Macmillan's anxiety must have been acute. Ever since 1946, when the United States had so abruptly ended the wartime atomic partnership, Britain had striven hard to renew it. During the past eleven years of atomic isolation, regaining Anglo–American interdependence had been a prime objective of Labour and Conservative governments alike, taking priority over relationships with European and Commonwealth friends and allies. There had been some promising developments but always British hopes had been disappointed. Some limited co-operation had been established, but it had fallen far short of the close partnership the British desired so ardently.

Then the British atomic weapons trials,[22] leading up to a successful thermonuclear test in 1957, had demonstrated Britain's nuclear competence and her value as a partner, and the United States' perception of the advantages of co-operation was sharpened when on 4 October the Soviet Union launched Sputnik, the world's first space satellite. The Americans were seriously alarmed by the implied intercontinental missile threat, and feared that the Soviet Union was over-taking the United States in advanced technology. The Prime Minister had written to the

President on 10 October, the very day of the Windscale accident: 'Has not the time come when we could go further towards pooling our efforts and decide how best to use them for our common good?'[23] Eisenhower promptly invited Macmillan to Washington for talks in late October about military co-operation in atomic energy, and on 23 October they had issued a joint Declaration of Common Purpose in which the President had committed himself to seek the amendment of the McMahon Act to permit 'close and fruitful collaboration of scientists and engineers of Great Britain, the United States, and other friendly countries'.[24]

With so great a prize almost within his grasp after so many years, small wonder if Macmillan was appalled at the prospect of losing it because of an embarrassing mishap at Windscale. October 1957 – with Sputnik, the Declaration of Common Purpose and the Windscale fire – was a momentous month. Macmillan had returned triumphant from Washington, via Ottawa, to find the Penney report awaiting him. How could it best be handled?

'WHAT DO WE *SAY*?'

Drafts flew between No.10 Downing Street, the Cabinet Office, the Atomic Energy Office[25] and the Authority as the White Paper began to take shape. Everyone concerned worked intensively to get comprehensive information about the accident into the public domain as expeditiously as possible. The White Paper appeared only three weeks after the Penney Inquiry began and less than a month after the accident itself (see Appendix I). The Penney Committee, asked to prepare an abridged version of its report, reduced it from 111 to 67 short paragraphs, by removing some technical detail and by omitting the conclusions and recommendations (which were absorbed into an introductory memorandum by the Prime Minister).

The abridged Penney report was included in the White Paper as Annex I ('the cause of the accident and the measures taken to deal with it') and Annex II ('the measures taken to deal with the consequences of the accident'). The other contents of the White Paper comprised the Prime Minister's memorandum, just mentioned; the MRC report on health aspects of the accident (Annex

III); a memorandum from the Authority chairman to the Prime Minister (Annex IV); a note on the Calder Hall and civil power reactors (Annex V); and particulars of the three Fleck Committees and six working parties (Annex VI).

A MATTER OF COLLECTIVE RESPONSIBILITY

The Prime Minister's memorandum[26] explained that the accident occurred while the pile was shut down for a Wigner release, described here (though not in Annex I or the Penney report itself) as a routine operation. The immediate cause of the accident was said (as in Annex I) to be the second nuclear heating, applied too soon and at too rapid a rate, which caused the failure of one or more cartridges in the pile whose contents then oxidised slowly, eventually leading to fire in the reactor. (It did not distinguish uranium or AM cartridges.)

The accident, the Prime Minister's memorandum went on,[27] was due partly to inadequacies in the instrumentation provided and partly to faults of judgement by the operating staff, these faults of judgement being themselves attributable to weaknesses of organisation. Annex I also commented on inadequate instrumentation;[28] the Penney report itself had had a good deal to say both about instrumentation and weaknesses of organisation, but neither referred to faults of judgement. This wording occurs in the memorandum to the Prime Minister from the Chairman (Annex IV).[29] It regretted the disturbance and anxiety caused to so many people, and accepted full responsibility for the accident. Commenting on the causes of the accident and the faults of judgement, it went on to say that, as the responsibility was primarily collective, the Authority was not taking disciplinary action against any individual. It concluded with an appreciation of the devotion to duty and the prompt and efficient actions of those concerned in dealing with the accident once it had been discovered.

There was no disposition to take disciplinary action and Plowden would have totally opposed it.[30] Penney had been clear that his inquiry was a fact-finding inquiry and he was not in the business of apportioning blame; if intended as a disciplinary board, his committee would have had to be constituted as such, with certain safeguards for witnesses. David Peirson, who had acted as secretary to the Penney committee, wrote[31] later that the committee was appointed as a technical committee to find out what happened. Its members, he recalled, were quite firm in their opinion that they were not appointed as a disciplinary inquiry; the two outside members had expressed the strong opinion that it would have been inappropriate for them to have been asked to serve on a disciplinary inquiry. At some vital points in the technical story the committee had had to establish what actions were taken by individual people, but had confined itself to doing so and had not probed further to ascertain why. In this report the Penney Committee had had to say what was done; in one or two instances, it thought, incorrectly. However, the witnesses had had no opportunity of knowing that their actions would be criticised, or of justifying them.

The wording that linked 'faults of judgement' directly to 'weaknesses of organisation', however illogical, was apparently carefully chosen with the object of exonerating the operating staff. But it failed in this intention; the White Paper seemed to point an accusing finger at certain avoidable mistakes by a very few, comparatively junior staff. Understandably when the White Paper appeared it caused much distress, exacerbated by the press coverage that followed. The Windscale staff felt deeply wronged, and Hinton too was outraged on their behalf.[32]

HEALTH AND SAFETY

To return to Annex III of the White Paper, the MRC's report on health and safety: Lord Hailsham received it and forwarded it to the Prime Minister on 6 November. The MRC committee[33] that had produced it so quickly was chaired by Sir Harold Himsworth, Secretary of the MRC, and its members included all those radiological experts who were already thoroughly familiar with Windscale and the Windscale accident. The report confirmed[34] in fuller detail their former conclusions; they were satisfied that it was in the highest degree unlikely that any harm had been done to the health of anybody, whether a worker in the Windscale plant or a member of the general public. (See Chapter 9 below for a comparison of the health assessments of the accident on different occasions between 1957 and 1988.)

Finally the White Paper presented, in Annex V, a reassuring

note[35] – unsigned but apparently by Cockcroft, Owen and Jukes – explaining why a similar accident could not occur in Magnox reactors, whether at Calder Hall and Chapelcross or at any of the power stations of the first civil nuclear power programme. Annex V outlined the important differences between the old Windscale piles, the new reactors at Calder Hall and the more advanced reactors being designed and built for the electricity industry. All were graphite moderated, but in the new reactors Wigner releases would be required much less frequently; because the minimum graphite temperature in normal operations would be higher than in the old piles, some annealing would be going on continuously and so stored energy in the graphite would accumulate much more slowly. All the evidence, according to Annex V, indicated that the new reactors would be able to run for about five years before a Wigner release was needed. When it became necessary it would not present any hazard, for several reasons.

First, the temperature safety margin during a Wigner release would be much larger; instead of air the coolant was carbon dioxide, which does not react with uranium below temperatures of 650°–700°C, whereas uranium oxidises in still air at 350°C. Second, in the new reactors the coolant would be circulated in a closed circuit, not discharged to the atmosphere through a high stack, and this feature would make it possible, by using the main cooling fans, to control closely the temperature generated in a Wigner release. Third, the new design of BCDG in the reactor core would be fixed, not moving, and it would be able to monitor each fuel channel continuously; therefore the problem caused by the jamming of the BCDG would not arise. Fourth, the Calder Hall fuel cartridges were of improved design with a lower failure rate than the Windscale cartridges, and there would be yet further improvements in the civil reactor fuel elements. Even if a fuel cartridge did fail it would be promptly detected and dealt with; the amount of radioactive material that could be released into the closed coolant circuit would be small, and the amount that could escape into the atmosphere from leakages in the circuit would be too slight to constitute any hazard. Finally, in the Calder Hall and civil power reactors the instrumentation and control systems – including the system for regulating the rate of pile power change – would be much improved.

GRAPHITE

The first point made in Annex V (about the behaviour of graphite in the new reactors) was eventually more than vindicated as no Wigner anneals were ever needed. But such confidence was not as yet supported by the research data available in 1957. In April 1957 the Harwell scientists were still uncertain about their estimates of Wigner energy accumulation in the Calder Hall and civil reactors;[36] they could not rule out the possibility of stored energy building up to a dangerous level but considered that an annealing process of the Windscale type would be unsafe. Even a year later,[37] Alan (now Sir Alan) Cottrell, of Harwell's Metallurgy Division, thought that – with design modifications, strictly limited operating conditions, and some anneals – it might be just possible to run the civil reactors safely for 20 years; otherwise he feared their safe working lifetime might be as short as 7–14 years, or even less. As it turned out this was a very pessimistic view, but he was understandably cautious in the absence of firm data.

How had it come about that there was still so much ignorance and uncertainty in 1957–58 about this basic component of the whole British nuclear project, including the ambitious 1955 civil programme? As we saw in Chapter 3, the main problems of graphite had been recognised very early. Two American scientists visiting Harwell in 1949, Dr Teller and Dr Failla, had warned that the most likely cause of a serious reactor accident, which could affect a very wide area, was the energy stored in irradiated graphite, especially if a fuel rod caught fire.[38] Risley was not represented at the meeting, and the warning does not appear to have been passed on, or taken very seriously; certainly Risley knew nothing of it.

However, a small team at Windscale had been carrying out research into stored energy in graphite ever since the early days of the piles[39] and in 1952 it pointed out the potential danger from graphite oxidation if large amounts of energy were released during an anneal. A Harwell team pursued joint studies with the Windscale team until the leader, Harold Sheard, departed for Canada in January 1955. As for Wigner anneals, Sheard thought it almost certain that they would be perfectly safe at graphite temperatures below 400°–500°C, and probably no danger would be involved up to 500°C. At one meeting in 1954[40] there was a reference to coolant air 'fanning the flames' if too high a temperature were reached, but the matter seems to have rested there until October 1957.

Risley and Harwell reports produced during 1954 showed that graphite exposed to irradiation became very much more reactive than unirradiated graphite; there was also some concern over the effects of possible contamination by salt of graphite in the piles. The increasing rate of oxidation with prolonged irradiation looked like being serious, especially as graphite weight loss would reduce the structural strength. These irradiation effects could be extrapolated over an operating lifetime but were still uncertain. Some American information on graphite was available to the British during the 1950s but it does not seem to have been particularly helpful, and when Davey and others attended a reactor safety conference in Chicago in 1956 he found that he could not arouse any interest in discussing graphite safety problems.[41]

There was keen competition between Windscale and Harwell for the very limited experimental space available as the two Windscale piles were the only reactors in Britain that could at that time provide the irradiation conditions required. Demands on them were heavy, and this fact severely limited graphite irradiation research. In mid-1955, at one of the informal co-ordination meetings which allocated R & D responsibility between the Industrial and Research Groups, it was agreed to concentrate the graphite studies at Windscale.[42] This happened gradually and the Harwell work ended in April 1957. To Risley's annoyance Dunworth and Fry, two senior Harwell scientists, were sceptical about the capacity of the small Windscale team to pursue the research effectively. The 1957 change may have been unfortunate, but for years before that the work on graphite had been fragmented and short of staff and facilities.

Then suddenly the Windscale accident turned a searchlight on graphite and revealed the precarious foundation of knowledge on which the whole gas-graphite reactor programme rested. Cockcroft himself, the Authority Member for Research, at a post-accident meeting at Risley on 4 November, acknowledged 'a grave mistake in the level of our work on Wigner effects in the past and indeed until now the tendency . . . to diminish the scale of work'.[43]

A crash programme of research was quickly prepared by Schonland (deputy director of Harwell), Kronberger (Risley's research director) and others, and was agreed by Cockcroft and Owen in mid-November.[44] The programme was accorded the highest priority. It was to be carried out by a unified, and much enlarged, team of scientists centred at Windscale, with an outstation at Harwell. The Harwell outstation staff would have to spend much time at Windscale, but would be given support by Harwell divisions (especially Theoretical Physics, Metallurgy and Electronics). An experimental design office at Risley under a senior engineer was to provide the special equipment. For the first six months the graphite project was to be headed by Alan Cottrell, until he left the Authority to become Goldsmith Professor of Metallurgy at Cambridge. The new complement was nineteen at Windscale and eighteen at Harwell, but the staff was never quite up to strength and Cottrell even had to fight to resist cuts threatened by a general economy squeeze.

The duties of the project included[45] consideration of all possible methods of reducing energy storage in graphite by changes in the design and/or operation of reactors such as the Calder Hall and CEA stations, and obtaining sufficient information on stored energy in graphite to ensure the safe running of graphite-moderated reactors now constructed and to permit the design of new reactors so that their safe life was a minimum of 20 years. (This was an arbitrary figure.)

'In my view', Sir William Penney told the newly appointed Cottrell, 'there is nothing more urgent for the AEA to do.'[46] The work was indeed crucial not only to the Authority's plutonium production reactors and the research reactors at Harwell, but also to the civil power programme. If modifications in design or operating methods were going to be needed there was little time to be lost and, for each power station, deadlines were set by which design modifications or revised operating procedures must be finalised in order to be effective. Reactors 2 and 3 at Chapelcross had sleeves incorporated in the channels to raise the graphite temperatures, but it was decided later that these would not be required for the civil power reactors.

The graphite project, led by Greenough of R & DB (W) after Cottrell left in 1958, solved many of the graphite problems – and all the urgent ones – before it was wound up in May 1960.[47] In his final report Greenough recommended continuing long-term research into some subjects that were still not well understood, such as graphite growth mechanisms and irradiation damage. He was satisfied, however, that the emphasis on Wigner energy in the R & D programme could be greatly diminished. The latest assessment of minimum temperatures (that is, the lowest temperatures at which the reactors could be operated for a working life-time without dangerous accumulations of Wigner energy in the graphite) were considerably lower than those assessed a year previously. On this basis all the CEGB reactors would be safe to run for 20 years at 75 per cent load factor, and Calder Hall for 18 years. He believed the figures to be reliable but thought that additional data in 2 years' time might warrant a revised assessment and perhaps permit a further reduction in minimum temperatures.

In the event, the Calder Hall and Chapelcross reactors have continued operating for over 30 years without Wigner anneals or graphite problems, and the Nuclear Installations Inspectorate (NII) has now (in 1990) given approval for a working life of 40 years for both stations.[48] The Magnox power reactors have certainly had some difficulties but not with their graphite moderators, and they have been safely operated without anneals. In retrospect the experience of the past 30 years has fully justified the optimism of the 1957 White Paper authors rather than the pessimism of Cottrell and his colleagues. But hindsight does not alter the facts of 1957; the problems of stored energy in graphite were then so serious, and the knowledge of graphite behaviour still so inadequate as far as the civil stations were concerned, that there were stronger grounds then for caution than for confidence.

FURTHER INQUIRIES

The Penney report, as we saw, had recommended a technical evaluation working party to make 'an urgent and thorough study of all the technical information to be derived from the accident', and two further internal reviews: of the IG's organisation and staffing, and the control of health and safety in the Authority. These recommendations[49] reappeared in the Chairman's memorandum (Annex IV) as a proposed evaluation by 'some independent person of standing' with 'experience of large-scale organisations operating processes involving hazardous materials'. The Prime Minister had already approached Sir Alexander Fleck, who by 5 November had accepted.[50] There were to be three committees under his chairmanship; the terms of reference and membership of all three were announced in the White Paper.

THE WHITE PAPER (CMND 302)

Much had happened in the few days since the Penney report reached Downing Street late on 28 October 1957. During the evening of 5 November Brook, Plowden and Bishop (the Prime Minister's private secretary) put the White Paper into final shape. They then suggested to the Prime Minister that he should tell the Cabinet how matters stood.[51] This he did next morning, and all was ready for publication.[52] Mr Macmillan made a statement to the House of Commons on 8 November, and the White Paper – 'Accident at Windscale No.1 Pile on 10 October 1957' – was published the same day, as Cmnd 302.

In the three to four weeks between the accident and the publication of the White Paper, press coverage had been fairly favourable, or at least neutral.[53] There had been some criticisms of apparent attempts to minimise the gravity of the situation, of inadequate district warnings, or lack of information, but the main criticism had been that, instead of a public judicial inquiry, the accident was being investigated by a board consisting entirely of Authority employees or ex-employees. *The Economist* had been particularly emphatic on this point.[54]

Press reactions after publication of the White Paper were gratifying to its authors and to the government. *New Scientist*, for example, which had earlier been critical, wrote on 14 November:

> The accident ... which resulted in radioactive iodine being released in small quantities over part of the Lake District caused much alarm. It could easily have resulted in opposition to the atomic power programme on which the country has embarked. But the official Windscale Report offers so frank and satisfying an explanation that the public is enabled to view the happenings of 10 October in [their] proper perspective.

The Prime Minister was pleased with the way it had all gone. 'I think that the Windscale announcement and the setting up of enquiry have been very well received', he wrote to Plowden[55] on 11 November. 'I am afraid that all this has caused you great anxiety but I want you to know that I have complete confidence in you and in your handling of this difficult matter. I shall continue to give you and the Authority all support during these trying days.' 'Thank you for your letter of 11 November and the kind things you say', Plowden replied. 'The knowledge that we had your support was a great help to all of us in dealing with the Windscale accident. We are most grateful.'[56]

Everyone seemed pleased except some of the Windscale staff and their staff association, the Institution of Professional Civil Servants (IPCS). They were troubled both by the White Paper – in which, they said, IPCS members were both implicitly and explicitly criticised – and by the consequent press coverage, which referred to 'a human misjudgment', 'an inexperienced operator' and men in charge who were 'not well qualified scientifically for the job'.[57]

The White Paper had not named names but from one paragraph a member of the Windscale staff was – incorrectly – identified in the *Daily Express* as the scientist who had made 'the initial mistake'. Whether or not a mistake had been made, the man named had not been on duty at the time. But he was pilloried in the press, and he and the IPCS complained about his treatment.[58] He received an apology from the newspaper. However, complaints about the White Paper itself – as distinct from the press coverage – were longer lasting and more substantial.

TELLING THE WORLD

This major accident, said *The Economist*,[59] was bound to be regarded with mingled interest and alarm by scientists the world over. The news was immediately picked up abroad from press and radio. The countries with most reason to be interested were the United States, Italy, Japan, Belgium and France.

The only country well qualified to advise and assist was the United States, with its unique atomic expertise and long experience of graphite moderated reactors, and the Americans, as we saw, had reacted very quickly. Dr Libby, the acting chairman of the USAEC, cabled Sir Edwin Plowden on 12 October with a generous offer of help.[60] He also wanted all the technical information that could be extracted from the accident, and the USAEC representative in London made another approach to the Authority after the press conference on the White Paper on 8 November.[61] At the beginning of December a team of five USAEC scientists and engineers came to England for a three-day conference at Risley.[62] The fifteen British participants were led by Cockcroft, Owen and Schonland, and the exchange of information – classified, significantly, as Secret Atomic – was extremely frank and free. The British gave their guests a full account of the accident, with all the relevant pile records. In return the Americans gave ample information about operating experience in their research reactor at Brookhaven, in New York State, and even in their big plutonium-producing reactors at Hanford, Washington State. The British had previously had some unclassified American reports on graphite, and in 1953 and 1956 had attended conferences in the United States on graphite and on reactor safety. They had been trying to arrange a further conference on graphite when the Windscale accident supervened.[63] But the Hanford reactors had always been covered by the strictest secrecy until these post-accident discussions, when they learnt more than had ever before seemed possible. The Americans gave details of Wigner problems, cartridge failures, malfunctions and accidents and described their procedures for carrying out anneals. A second conference was held in January 1958 when nine American scientists came to England for comprehensive discussions on the health physics and medical aspects of the accident.[64]

In Europe the Italian government had been prompt in asking for information, and had approached the British Embassy in Rome on 12 October.[65] The Italians were in the process of buying Magnox reactors from Britain for their first nuclear power station, at Latina. The reassurances relayed from London to Rome were clearly satisfactory as the Latina contract was signed in November. (The Japanese, too, were buying a Magnox power station, Tokai-mura; this and Latina are the only nuclear power stations Britain ever exported.)

The Belgians had a small graphite-moderated research reactor at Mol. It was similar to BEPO, at Harwell,[66] and after the Windscale accident they asked for advice about Wigner energy.

The Mol reactor was not expected to require annealing before the end of 1959, and the Belgians were invited to send scientists to Harwell to witness the next anneal of BEPO in the spring of 1958.

Next to Britain, the European country with the greatest interest in graphite reactors was France. The French Commissariat à l'Energie Atomique (CEA) had a graphite-moderated reactor (air-cooled) at Marcoule, and was building two more (to be cooled by carbon dioxide). They might have been expected to show a special concern about what had happened at Windscale but, surprisingly, by the end of November the French authorities had still not asked for information about the accident. They may perhaps have held back because they anticipated a refusal, especially as Windscale was a defence plant. Officially Anglo–French atomic relations were not close because Britain's existing obligations to the United States and her hopes of a renewed Anglo–American partnership over-rode all other considerations; however, personal relations between Authority and French CEA scientists were excellent. The British decided to take the initiative and invited the French to unclassified talks in February. They gave their visitors an account of the accident, and discussed with them its health and safety aspects, the need for pre-determined maximum permissible levels of radiation in emergencies, and various administrative problems.[67] The talks appeared to be very successful and were much appreciated by the French participants.

At the time of the accident, no warning had been given to other countries of a possible airborne radiation hazard. This was not strange. The environmental hazard was believed to be very limited in extent, affecting only a 200 square mile area around Windscale, and that hazard had been controlled by imposing a milk ban. The fallout from Windscale was not thought to pose a significant public health hazard in most of Britain, and certainly not to other countries. Such international machinery as now exists for exchanging information about nuclear accidents was not then in operation. The Euratom Treaty had been signed only in March 1957, and Britain was not a member. The International Atomic Energy Agency in Vienna (IAEA) had just come into existence in June 1957. The European Nuclear Energy Agency (ENEA–now the NEA) was not created until 1958. However there was close co-operation at the working level between Authority health physicists and overseas scientists. Increased radioactivity – including polonium-210 and tritium – was detected in parts of northern Europe within a day or two of the accident. Harwell scientists asked for air filter samples from laboratories in Europe, including Copenhagen, the Hague, Bonn and Oslo and, as it was International Geophysical Year (IGY), many samples from abroad were obtained with the help of the IGY Advisory Committee on Nuclear Radiation.[68]

7 Three More White Papers

THE FIRST FLECK REPORT, ON THE ORGANISATION OF CERTAIN PARTS OF THE UKAEA (CMND 338)

Of the Fleck inquiries set up after the original November 1957 White Paper (Cmnd 302), the first to submit its recommendations to the Prime Minister was that on organisation. The report – prepared by Sir Alexander Fleck,[1] Mr C. F. Kearton[2] and Sir William Penney – was submitted and published in December 1957 as Cmnd 338.

It dealt mainly but not exclusively with the IG. It gave a brief historical account of the development of the Authority, and particularly the IG, from its origins in 1946 in the Ministry of Supply.[3] It described the initial task of producing plutonium for military use; and the 'extraordinarily rapid and successful development in an entirely new field of technology' by the Group. It itemised the tremendous workload that the Group had had to carry: in meeting the demands, first, of the expanding military programme (though, oddly, it did not mention H-bomb requirements); second, of the civil nuclear power programme; and, third, of the development of other more advanced reactor systems.[4] The report paid tribute to the 'extraordinary degree of vitality and efficiency with which the Industrial Group staff [had] carried out the responsibilities that [had] been laid upon them';[5] all this had been done, and the organisation expanded, in the face of a serious national shortage of technical manpower. The main conclusion of the report was that weaknesses in the organisation of the IG stemmed 'primarily and noticeably from an insufficiency of senior technical staff in the Operations Branch'. It was more difficult for the IG than for Harwell to attract sufficient well-qualified recruits; within the IG the Operations Branch found it more difficult than the Engineering and R & D Branches, which seemed more exciting and glamorous. The Operations Branch was the atomic Cinderella.[6]

At this time the IG's total strength[7] was just under 15 000, of whom 1200 (8 per cent) were scientifically or professionally qualified. At Group HQ at Risley, the non-industrial staff of about 2500 was organised in eight branches – Engineering, R & D, Technical Policy, Industrial Power, Operations, Health & Safety, Administration, and Accounts & Stores – each under a director; the directors, with the managing director as chairman, constituted the Production Executive Committee (PEC). The managing director had no deputies; above him, the all-important post of Authority Member for Engineering and Production had been vacant since the departure of Sir Christopher Hinton in August 1957.[8]

Of the five IG establishments, the Fleck Committee was concerned with Windscale and Calder, Chapelcross and Dounreay. At this time Dounreay was a new site, not yet operational, and was run from Risley by the Director of Operations. Windscale, Calder Hall and Chapelcross were all under a single works general manager, and he was very short of senior staff.[9]

The White Paper made some general Authority-wide comments before proceeding to particulars. As a major cause of recruitment problems and staff shortages it pointed to the government decision, announced at the same time as the decision to set up the Authority, that the salaries of senior staff 'should not be seriously out of scale with those paid by other public corporations', and that below this level the Authority should inherit civil service grading structures and salary scales.[10] Another comment of Authority-wide application[11] was that too much reliance had been placed on committees as a substitute for executive action, and the Fleck report recommended that the Authority's entire committee structure should be reviewed.

For the IG, Fleck said, the most important priority must be to show that the large-scale plant it had in operation could be worked in an efficient and safe manner.[12] In the recent past perhaps too large a proportion of the limited technical resources of the group had been deployed on studies of advanced types of reactor. As to the Group's structure,[13] the committee considered that it was an essential unity and that its functions should not be divided.[14] Instead, the management should be reinforced by providing the managing director with two deputies and a non-executive director. Generally, because of the highly decentralised nature of the Authority – and perhaps also because of staff shortages and time pressures – there had been insufficient liaison with other parts of the Authority; in future one member of the Group management board should be specifically charged to ensure effective liaison and to see that all the technical resources of the Authority were available to the Group. Representatives of other groups and of the London office should be visiting members of the board.

Within the Group there was some imbalance between branches,[15] partly due to the sheer pace of change. The Engineering Branch was very large but its organisation was satisfactory. The Industrial Power Branch, a recent formation and still rather small, was of the utmost importance; it needed to be expanded and strengthened at once, so that it would be fully equipped to carry an increasing workload and meet its responsibilities to industry and the electricity authorities for the civil nuclear power programme. R & DB was also well organised but, while the Operations Branch was facing so many serious problems, R & DB was devoting too much of its effort to the study of future systems not yet authorised for construction, or even ready for full design study. R & DB, it is true, had set up sections specifically to undertake work for Operations Branch, but they were under-used because Operations Branch lacked the technical staff to take the initiative in formulating the problems needing attention.

The Technical Policy Branch, Fleck said, should transfer some of its responsibilities (e.g., for commercial policy) to the Authority's London office and should then be able to concentrate on forward planning and study of alternative policies. The Director of Administration at Risley ought to become Director of Personnel and Administration and should be enabled to play a full part in the development of personnel policy in the Authority as a whole; his branch should be very considerably strengthened to deal effectively with the crucial tasks of recruitment, deployment, training and career planning. The Group's Health & Safety Branch was mentioned in passing, but the important subject of health and safety was about to have a White Paper of its own.

The report's special concern was with the Group's Operations Branch[16] and the factories. It recommended that the Director of Operations at Risley should have two deputies to assist him: a deputy director (production) and a deputy director (technical). The former would have oversight of the works general managers and their factories; the latter would have under him a chief chemist, a chief engineer, a chief metallurgist and a chief physicist, each with his technical section. They would be responsible for liaison with the technical staff at the works; for keeping a close scientific watch on pile loading; for checking operational manuals; for commissioning new plant; and for operational research.

As to the factories,[17] the main Fleck recommendations were that a separate works general manager should be appointed for Chapelcross, reporting directly to the Operations Branch at Risley. Each works general manager should have a deputy, and a works engineer, a health and safety manager and a works secretary reporting to the WGM. He ought also to have more technical support, with scientific and technical staff responsible to him through a non-executive technical officer.

The status of operational posts throughout the Group needed raising, but the committee believed that, with the right leadership, the business of plant operation could be made attractive to highly qualified men, especially if interchanges of staff were encouraged between the Operations, R & D and Engineering branches, and with other groups of the Authority. Senior Operations Branch posts needed to be filled by men with good technical abilities as well as management skills.

There were special problems at Dounreay,[18] at this time directly under the Director of Operations at Risley. The report said that a new post of Director of Dounreay should be created, with a seat on the Group board of management, and a man of high scientific competence must be chosen. Meanwhile the Authority should 'carefully review the time scale and staffing problems of the Dounreay project in the light of its other commitments'.[19] Though it was not within its terms of reference, the committee felt moved to comment also on the current proposals to set up a new Research Group establishment at Winfrith Heath in Dorset. 'The same requirements for technical staff of the highest calibre and in considerable numbers will obtain', it said, 'if advanced reactor systems are to be set up on this site.'

To a large extent, the White Paper put into cogent and incontrovertible form what Hinton and Owen had both been saying for years, and had reiterated in evidence to the Fleck Committee. It might perhaps be inferred that the IG had for years been failing to recruit staff, but this was not so. In two years the Group had dealt with 28 000 responses to job advertisements and had interviewed 6000 people. Some staff spent a large part of their time interviewing (at one stage Owen did nothing else for six months). Between August 1954, when the Authority came into being, and October 1957, the IG's non-industrial staff had increased by over 69 per cent from 3364 to 6166.[20] For Windscale, Calder Hall and Chapelcross combined, the increase was 22 per cent, from 979 to 1195, but against the approved complement of 1320 there was a 9.5 per cent shortfall.

With such a massive expansion in very little more than three years, why were there such serious staff shortages, and such complaints about recruitment? There were three reasons. First was the failure to recruit key senior staff, and staff in grades and classes of which there was a shortage throughout the country. The second reason was that a significant part of the increase did not represent real gain, but merely the replacements needed when the Authority, on its formation, lost the services hitherto provided by the Ministry of Supply and the Ministry of Works, and so had to build up its own infrastructure. The third reason was that, however fast the staff might grow, the Authority's commitments (both the defence commitment and the recently initiated civil programme) grew and multiplied much faster, with the inevitable mismatch described in Chapter 3.

Sir Leonard Owen was both pleased by the Fleck Report (since it largely embodied and confirmed his own views) and exasperated. He pencilled these hasty notes[21] in the file:

> From the public presentation point of view I would accept this report *in toto*, with perhaps a word about the fight we have continuously made to get adequate staff. (The report almost sounds as if we didn't recognise the inadequacy). There is a sense of unreality in the phrasing that leaves me gasping. We have fought and scrapped during 11 years to get and keep staff... Impossible conditions in the M of S [Ministry of Supply] for getting top people was one of the cogent reasons for setting up the AEA. In spite of all this our top structure has never been adequate. Are we going to get a new magic wand...?
>
> The report misses one cycological [*sic*] aspect and that is that the IG has always put as the paramount issue the necessity of *meeting the Defence Programme. Shortages of men or materials or knowledge were not allowed to jeopardise this. The times given were such that risks had to be accepted.* (My emphasis)

The Authority[22] studied the Fleck report and in January Plowden wrote to the Prime Minister to say that the Authority endorsed, and would follow, the general principles of organisation laid down and the general scheme recommended for the IG. Over a period of time the committee's recommendations would be implemented as nearly as possible, and every effort would be made to fill vacancies either from within the Authority, or from outside recruitment.

The IG was going to need 60–70 new senior posts in addition to 60 existing vacancies. That meant recruiting 120–130 senior staff, and the aim was to fill the key posts in six months and the remainder over about three years.[23] The new Member for Engineering and Production, Sir William Cook[24] (appointed in January 1958 after a five-month gap), and Owen, the Managing Director, faced a daunting task; so did the newly uprated Personnel and Administration Branch. But the White Paper provided them with the magic wand previously denied to Hinton and Owen. By early 1959 the IG was almost up to 'Fleck' strength:[25] that is, complements much above the pre-accident complements (themselves never filled) which, on the clear evidence of the Fleck Committee, the Authority and the government now accepted to be seriously inadequate. The Authority's total strength[26] rose from some 27 000 in 1957 to 38 500 in 1960, and by 1962 to over 40 000.

The recommendations about Windscale, Chapelcross and Dounreay were all carried out, the Dounreay programme was slowed down as the White Paper had advised,[27] and the completion of the DFR was deliberately delayed. But the hint about Winfrith was not taken and the new experimental reactor site there went ahead;[28] the first buildings there were occupied in March 1959, with a Research Group staff on the site totalling 750.

As for the structure of the IG, there was soon an important departure from the Fleck White Paper. Fleck, it may be recalled, had considered the IG an essential unity and had recommended that it should not be split (a possibility that had been considered earlier by the Group management). Sir Leonard Owen reviewed the position in January 1959.[29] During 1958 an all-out effort had been made to meet the Fleck recommendations on organisation. Recruitment had been reasonably successful up to a point, and staff in the IG had been redeployed, largely from headquarters at

Risley to the factories, to the extent that important posts in the Industrial Power Branch were vacant and the Operations Branch was virtually denuded; not what Fleck would have wanted. Even so, 50 per cent of the Group's health and safety posts remained unfilled. Moreover, no additional strength in the top management had been obtained from outside the Authority, and the management load was still excessive, and bound to grow over the next few years. (The formidable span of control is apparent from Table B of Appendix II.) All this, in addition to the special problems of this new, complex and rapidly advancing technology, convinced Owen that a single Group was not practicable. The Authority agreed[30] and in March 1959 the IG was divided into two; a Production Group (under Sir Leonard Owen, as managing director and Authority Member for Production), and a Development and Engineering Division (under Sir William Cook). Lord Fleck sent a strong protest to Sir Edwin Plowden about the 1959 split, but the authority believed that the size and diversity of the IG's activities made the arguments overwhelming. The subsequent history of the reorganisations of the northern groups, and later of the whole Authority, are outside the scope of this book, but the Windscale accident and the first and second Fleck reports had great influence on the Authority's structure and staffing for years.

THE SECOND FLECK REPORT, ON THE ORGANISATION FOR CONTROL OF HEALTH AND SAFETY IN THE UKAEA (CMND 342)

The second of the Fleck reports was submitted to the Prime Minister on 20 December 1957 and was published in January 1958 as Cmnd 342. It dealt with the health and safety organisation in the Authority – Hinton regretted that its terms of reference were not wider[31] – and was produced by a committee on which Fleck, Penney and Kearton were joined by the Chief Inspector of Factories, the Chief Alkali Inspector, and an eminent radiologist, Professor B. W. (now Sir Brian) Windeyer.

As the Penney report had shown (and as Chapter 4 suggests) the Authority's health and safety organisation was weak and poorly co-ordinated, and its responsibilities were inadequately defined. This had long been realised by Andrew McLean, the principal medical officer at Risley, who had been thinking about it deeply and working out his ideas. Between the accident and the time when the Fleck Committee began work, the Authority had acted rapidly to strengthen the organisation,[32] and the Fleck report discusses both the earlier and the improved situation before making recommendations for further improvements.

The health and safety organisation put in place when the Authority was created in 1954 had consisted of health physics departments and medical departments at each Authority site, responsible to the works general manager or site director. There were no health and safety staff in the London office and no central body concerned with health and safety. The Authority Member responsible for radiological health and safety throughout the Authority was Sir John Cockcroft, as Member for Research, but he had many other responsibilities and no supporting staff (other than the medical and health physics staff at Harwell, who had no defined role outside the Research Group). Reactor and plant safety was the responsibility, at board level, of the Member for Engineering and Production. The latter had created a small safety branch at Risley towards the end of 1956.

Authority-wide co-ordination was effected by a network of committees. There was a high-level committee set up in 1954, the Project Health Committee chaired by Plowden (later by Cockcroft) and attended by the radiological consultants, which met twice a year. It had sub-committees on reactor operations, safety research, and reactor location; a Project Health Discussion Group (later called the Health Panel); and a committee of principal medical officers. Each of the three Authority groups – Research, Industrial and Weapons – also had a network of committees on health and safety matters. The committees were all advisory and the sole executive responsibility for health and safety rested with the works general managers (or site directors). The scope for divergent standards and practices and for conflicts of interest seems obvious, yet the Authority had extraordinarily wide corporate responsibilities for health and safety[33] under the statute which had created it in 1954.

Plans for more extensive machinery, and a permanent staff, were put forward in August 1956, and in January 1957 a new health and safety organisation for the Authority (but excluding the Weapons Group) was announced.[34] The Project Health Committee was replaced by a Health Advisory Committee chaired by Cockcroft as Member for Scientific Research; it included the Authority's medical and biological consultants as well as the chief medical officer and chief health physicists, and was to advise the AEX. The Health Panel was to continue. Another new committee, also reporting to the AEX, was the Safety Executive Committee, chaired by the Member for Engineering and Production, which was to apply the standards recommended by the Health Advisory Committee; the specialist subcommittees of the old Project Health Committee were transferred to the Safety Executive Committee. It was agreed that the Weapons Group would remain largely outside the new structure. The Chief Safety Officer and his branch being formed at Risley would serve the AEA as a whole.

These new arrangements were again in the melting pot even before the Windscale accident. During the summer of 1957 it became clear that greater integration was needed between the health aspects and safety aspects of work in the IG. New plans were put forward in September and were implemented in October, shortly after the accident. They involved yet further changes in the committee structure. There was now to be a top-level Executive Committee on Health and Safety, chaired by Sir John Cockcroft and with three subordinate Committees: the Health Executive Committee (also chaired by Cockcroft), the Safety Executive Committee (under Owen) and the Weapons Safety Executive Committee (under Penney). In 1958 the name of the Executive Committee on Health and Safety was changed to 'Authority Committee on Health and Safety', and the subordinate committees were also renamed.

The October reorganisation also involved the strengthening of the Risley health and safety organisation in a new branch, with McLean as Director of Health and Safety. The branch consisted of two divisions. The first was under F. R. Farmer, who had been appointed Chief Safety Officer in 1956. His small staff, as we saw, was still much below strength in October 1957, and was fully occupied with work on new plant and future systems. His division was to have three safety officers dealing with engineering safety, health physics and criticality problems. While having a functional responsibility for the health and safety teams in the IG factories, it was intended also to have an Authority-wide responsibility for reactor and plant safety and criticality. The second division was the medical division, which was to be concerned only with the IG and its medical and nursing staff.

Many factors militated against a strong Authority-wide health and safety organisation. There was a shortage of experienced and well-qualified health and safety staff, and they were urgently needed at the sites. Then there was the decentralised character of the Authority; the Groups had considerable autonomy and the sites too were highly independent, even isolationist. With all this, and the geographical distances involved, the Authority was not an easy organisation to control from the centre.

The reorganisation of October 1957, as appeared later, had not gone far enough. But it was a move in the right direction and the Fleck report[35] endorsed the new committee structure and the new health and safety organisation at Risley (and the concept of an Authority Safety Branch which it contained). The report recommended that the latter should be brought up to strength as a matter of urgency. The new Health and Safety Branch at Risley, the report said, should have wide-ranging terms of reference, including the preparation of codes of practice for the safe design and operation of reactors, chemical plant and laboratories; an inspection service to ensure compliance with the codes of practice; and the provision of radiological and medical services at each site. It should be concerned with the detailed application of the standards of radiological protection laid down by the MRC for radiation workers and the general public. It should also provide an advisory service for the electricity authorities, and should liaise with government departments and with other countries.

The report endorsed the committee structure recently established as providing an effective means for overall supervision of health and safety. It advised that in each of the IG factories the management of health and safety should be integrated into a single health physics/medical/safety branch responsible directly to the works general manager. It recommended that the site emergency plans and procedures at each site should be brought to a uniformly high standard and co-ordinated to ensure that the Authority's maximum technical resources could be quickly deployed to meet an emergency at any site.

In the eighteen months after Cmnd 342, the health and safety organisation underwent several more changes.[36] The concept of an Authority Health and Safety Branch (AHSB) approved by the Fleck report was developed further, and the compromise arrangements described above were replaced by an AHSB, independent of the Groups, and with genuinely Authority-wide responsibilities; to demonstrate this McLean, as Director of Health and Safety, moved from Risley to London in 1959. The new branch consisted of three divisions. The Radiological Protection Division (RPD), headed by Marley, was located at Harwell, and so was the Chief Medical Officer. The RPD drew its staff mainly from Harwell, but the site necessarily retained separate health and safety and medical staff of its own. The second AHSB division, under Farmer, was called the Safeguards Division;[37] it was located at Risley (later, at nearby Culcheth) and it dealt with all aspects of reactor and plant safety including criticality. The third AHSB division was a small administrative unit in the London office.

With no executive powers the AHSB had to be both diplomatic and determined in order to establish itself, gain confidence, win inter-Group co-operation, change some entrenched isolationist habits (for example, in such matters as standardising health statistics) and make functional lines of communication strong and effective. It succeeded in the next ten years or so in preparing many health and safety codes, raising health and safety standards and co-ordinating practice within the Authority, and participating strongly in national and international policy making. In ICRP, the United Nations Scientific Committee on Effects of Atomic Radiation (UNSCEAR), the IAEA and the ENEA it made important contributions to discussions of radiological protection and reactor and plant safety.

One Fleck Committee recommendation[38] went well beyond its terms of reference. It had become apparent to the committee how very few experts there were in the radiological health and safety field. Throughout the AEA there were only about 90 staff with honours degrees or equivalent qualifications who were engaged full-time on health and safety duties. The workload of the health and safety staff (especially of Marley, the chief radiological safety officer at Harwell) was excessively heavy because of external demands: to contribute to ICRP, UNSCEAR and international conferences, to sit on committees and to advise government departments. Such demands were bound to grow as the uses of atomic energy increased, and specialist staff would also be needed by the electricity authorities and the public services. The Fleck report recommended that the Authority should take the lead in building up the country's supply of specialist medical staff and should set up a national training centre for health physicists and nuclear safety staff with courses graduated to suit all levels.

Hinton, as we saw, had regretted that the Fleck Committee's terms of reference were limited to the Authority's health and safety arrangements. The matter, he thought, should be considered on a broader basis, because there would soon be a considerable number of nuclear power stations around the country, and because the public should be assured that atomic health and safety were not under the same management as operations, but were supervised by an independent nuclear safety body. The Authority, he said, had considered such an idea early in 1957, and had concluded that some such development was inevitable.

This idea was already under consideration by an interdepartmental committee[39] as a result of Hinton's and the Authority's suggestions, and the outcome in 1959 was the Nuclear Installations (Licensing and Insurance) Act which, *inter alia*, set up the NII. The Authority was to play a key role in the establishment of this body – by secondments and transfers of personnel, helping in recruitment, training staff (at the Calder Operations School and by attachments), and assisting the Inspectorate – until it was able (from 1963 onwards) to carry out its duties itself. The Authority itself was not then subject to inspection or licensing by the NII, but has recently become so.

It is sometimes believed that the 1959 Act and the NII resulted from the Windscale accident. This was not so, although the accident undoubtedly increased the government's sense of urgency and the Fleck committee gave strong support to the idea. Incidentally, the interdepartmental committee that advised on the setting-up of the NII gave much thought to where it should be located in the machinery of government. It was anxious to ensure its independence and to avoid the conflict of interest that could result if it was placed in a ministry responsible for atomic energy promotion or operations. The committee therefore advised against putting it in the Ministry of Fuel and Power, but Ministers decided otherwise. Today it is in a neutral position under an independent statutory body, the Health and Safety Commission.

One new body whose origins can certainly be traced to the Windscale accident and the Fleck report is the National Radiological Protection Board (NRPB). The Fleck recommendation that the Authority should set up a national training centre was remitted to another committee (under the late Sir Douglas Veale) for study. Its 1960 report[40] did not favour the training centre idea but it strongly recommended a separate national body, outside the Authority, to give independent advice on radiological health and safety matters. Its purpose would be to provide a demonstrably independent source of advice and information, not subject to possible conflicts of interest; to reduce the overload on the Authority's own health and safety specialists, who had ample responsibilities of their own; and to make the most economical use of scarce skills.

The Veale recommendation of an independent national body was endlessly discussed in Whitehall, with much argument over the cost. But eventually, in 1970, it reached the statute book when the Radiological Protection Act set up the NRPB.[41] Its first director was Andrew McLean; after his untimely death in 1981, he was succeeded by John Dunster.

The creation of the NRPB in 1970, as recommended by the Veale report in 1960, altered the position of AHSB. So did the hiving off in 1971 of the Authority's Production Group to become British Nuclear Fuels Ltd (BNFL), which removed from the Authority the industrial sites with the biggest radiological protection workload. Many of the RPD personnel were transferred to the new NRPB, joining others from the MRC's Radiological Protection Service to form the initial staff of the Board. The Safeguards Division of AHSB, headed by Farmer, remained within the Authority as the Safety and Reliability Directorate (SRD). It continues to apply its accumulated expertise to nuclear and non-nuclear safety problems in the Authority and, increasingly, for outside clients.

THE THIRD FLECK REPORT, ON TECHNICAL EVALUATION (CMND 471)

The third Fleck Committee was appointed 'to make a technical evaluation of information relating to the design and operation of the Windscale piles, and to review the factors involved in the controlled release of Wigner energy'. Its members were Fleck, Cockcroft, Penney, Diamond and Kay, with the physicist Professor H. W. B. Skinner of Liverpool University (formerly of Harwell) and the Harwell chief chemist, Robert Spence.

The Technical Evaluation Committee (TEC) worked mainly through six expert working parties: on graphite, cartridges, instrumentation, filters, Wigner releases and operational problems. These absorbed the working parties that Owen had set up earlier (Chapter 5).

An interim TEC report[42] on 21 December 1957 dealt mainly with arrangements for a planned Wigner release in BEPO, Harwell's first experimental reactor. As for Windscale Pile No.2, on existing evidence it believed that – provided certain alterations were carried out – there was little doubt that it could be operated safely (but see below). Plowden advised against publishing the interim report and so reminding the public of the Windscale accident again; the Prime Minister agreed. It was in any case of little general interest. The final TEC report was submitted to the Prime Minister in June 1958 and published in July as Cmnd 471. The separate working party reports were not published, but the working party chairmen joined the TEC members in signing the final report, to indicate that in its analysis the Committee had correctly interpreted their views.

Inevitably the TEC worked over a great deal of the same ground as the Penney Inquiry, and with the benefit of all the additional experimental work and calculation which delayed its report by three months beyond its planned completion date. However, the TEC was charged to look to the future, rather than the past; it was not to concern itself with the cause of the fire, or to question the finality of Cmnd 302 on that score. This restriction must have been difficult for the TEC since to know exactly what had happened in Pile No.1 was surely relevant to its chief concern, the future safe operation of Pile No.2.

Before considering the TEC report itself, we may look at the background to the Committee's work: the working party discussions and reports. Here the work of the TEC, the activities of the IPCS and the views of the staff with most experience of the piles themselves become intertwined, but for the sake of clarity they are unravelled here and the IPCS aspect will be dealt with later in this chapter.

Filter Working Party

Before turning to the TEC final report, the TEC working party papers and reports merit attention as they are more detailed and informative. The filter working party[43] – after considering the history and design of the filters, the course of events up to 12 October, and the results of radiological surveys and measurements – came to some important conclusions, both about the accident to Pile No.1 and about the safe operation of Pile No.2 if restarted. It judged that in the past the filters had been unsatisfactory (see Chapter 3), even in the non-accident conditions for which they had been designed. With the significant improvements made or in hand, they would still be inadequate if Pile No.2 suffered an accident comparable to that in Pile No.1; if several tons of uranium were to be oxidised in an accident, a filter efficiency of 99.8 per cent for iodine-131 would be necessary to prevent a district hazard.

The working party concluded that it was not the filters themselves which had (as Cmnd 302 said) prevented the escape of most of the radioactive material from the damaged core of the pile. Of the uranium burnt, and the fission products it contained, only a small fraction reached the filters at all. In the very high temperatures of the fire, a compact and brittle crust had formed on a layer of oxide, which contained the molten metal and trapped most of the non-volatile fission products. But a large fraction of volatile iodine-131 escaped from the core, much of it passing through the filters; this explained the disproportionate release of iodine-131. Of the estimated 70 000Ci of iodine-131 contained in the oxidised fuel, the working party believed that 50 000Ci reached the filters; they retained 30 000Ci, adsorbed on particles of lead and bismuth oxides from the isotope channels, rather than free. Some 20 000Ci were emitted to the atmosphere. The efficiency of the filters for iodine-131 therefore – even helped by the adsorption factor – was only 60 per cent. (For later estimates of the fission product inventory and the destination of the various radioactive materials, see Chapter 9 and Appendix IX.)

No standard had yet been set by the MRC, but it was assumed that the tolerable emission of iodine-131 in an isolated incident would be 100Ci (and for radiostrontium 10Ci). Therefore in any similar accident in Pile No.2 a filter efficiency of 99.8 per cent would be required to prevent a district hazard even if most of the iodine-131 was not free but was associated with particulate matter, and this might well not be the case.

The existing filters, with an estimated 60 per cent efficiency, would not (the working party said) be able to cope safely with any accident in Pile No.2 in which more than 30kg of uranium fuel was burnt. Without a guarantee that a fire occurring during a Wigner anneal could be confined to 30kg of fuel, some further means must be found of preventing the release of radioactivity. The filters might achieve an efficiency of 85 per cent with improved oiling techniques, and this could be enhanced if a greater pressure drop through the filters — entailing some loss in production — could be allowed. The next possibility was the installation of a chemical scrubber by-pass system for decontaminating the coolant air before discharge. (A revised assessment made in 1981 will be discussed in Chapter 9.)

Graphite

The graphite working party[44] studied data from the United States and the USSR as well as Britain. It noted that prolonged irradiation of graphite greatly increased the rate of oxidation, and that oxidation was significant at 450°C although not at 350°C and below. It concluded that if a further Wigner release were carried out in Pile No.2 by the usual method, temperatures of over 400°C must be expected, possibly 500°C in some small regions. At temperatures above 450°C, oxidation of the graphite might produce heat at a rate that the pile blowers could not cope with; it was therefore unsafe to carry out a further release in Pile No.2 by existing methods. Completely safe operation could best be achieved by annealing the graphite at nine-monthly intervals by blowing electrically heated air through the pile, not by nuclear heating, but this might prove impracticable or excessively expensive. If so the possibility of operating the pile for a further (unspecified) period without annealing should be considered.

Instrumentation

The instrumentation working party[45] thought that the existing instrumentation had been adequate by the standards obtaining when it was installed, but could not be considered so in 1957. It recommended that to measure fuel temperatures there should be 100 thermocouples during normal operations and 300 during Wigner releases. For graphite temperatures, there should be 300 at all times. With these numbers involved a digital system of data processing and recording would be essential, allowing a complete scan of 600 thermocouples in about ten minutes. Individual warning levels could be set to give an alarm, and there should be at least two high-temperature trips. Other improved instrumentation was needed for measuring neutron flux and for detecting fission product contamination and failed cartridges at all times and in all modes of pile operation. Television apparatus was also required to enable the scanner gear and the whole rear space of the reactor to be visually inspected.

Cartridges

The cartridge working party[46] studied the uranium and AM cartridges; other isotope cartridges caused it little concern. It observed that for uranium fuel elements there was no maximum safe temperature *per se*; catastrophic oxidation could be initiated at low temperatures in some heat transfer conditions and safety therefore depended on being able to detect and discharge an oxidising element, in all modes of reactor operation, before the burning spread. In normal operation, with the blowers on, a maximum fuel temperature of 395°C was safe, but it would be potentially unsafe with the reactor shut down and the BCDG not working. Some reliable method for detection of failed cartridges must be available during shutdown.

In AM cartridges compatibility of the alloy and aluminium was a serious problem, and diffusion of lithium into the aluminium had been observed at a temperature of 250°C. When temperatures over 437°C were reached the aluminium might be completely penetrated in an hour or less, and at temperatures as low as 400°C bare lithium-magnesium alloy would oxidise catastrophically in air. The Mark III cartridge was therefore a hazard, and if it ignited the fire might spread to other cartridges or to the graphite moderator. To eliminate the possibility would impose such severe restrictions on pile operation that this design of cartridge should not be used in future. Some equipment similar to that for detecting failed uranium cartridges – the BCDG – was also required to detect failure of AM cartridges of whatever design.

Though the working party expressed no views on the cause of the fire in Pile No.1, its comments on the uranium and AM cartridges were suggestive. However, investigations of Mark III

AM cartridges[47] which had suffered damage in the piles (reported in an IG scientific report) identified one cartridge from Pile No.2 that had gone through the July 1957 anneal; the nearest graphite thermocouple, only 3ft away, had registered only 240°C. (Does this suggest that the AM cartridge failed at a very low temperature, or that the heat from the damaged cartridge did not spread in the surrounding graphite?) Examination of the damaged AM cartridges showed that there had been molten AM metal in the pile which had not ignited; this supported the view that an AM cartridge had not started the Pile No.1 fire.

The draft report submitted to the TEC, containing much of the material in the working party reports, aroused some strong criticism[48] from William Strath, Authority Member responsible for external relations:

> The Report conveys the impression that the Penney committee reached its conclusions over-hastily and on inadequate technical evidence. This is the combined effect of . . . paragraphs dealing with the lithium/magnesium cartridges, and various statements throughout the report about the lack of knowledge and theoretical understanding, which is still incomplete, of the behaviour of graphite under conditions of irradiation. The IPCS may claim on the basis of the Report that the Authority had insufficient grounds for stating that errors of judgment were committed by the operating staff. [One paragraph] suggests that the Authority were much more at fault in not recognising the hazards of the lithium cartridges . . . and in not taking adequate precautions to test them . . .
>
> Moreover [one sentence] makes questionable the part played by the second nuclear heating with which the 'errors of judgment' were linked. The Report leaves a big question mark over the behaviour of the Calder reactors and the civil power reactors . . . The emphasis placed on the limitations of our knowledge of graphite behaviour . . . strengthen [sic] this impression.

The final report published in July 1958 as Cmnd 471 (the fourth White Paper following the Windscale accident) endorsed the TEC recommendations about the Wigner release in BEPO which had been successfully carried out in March 1958.[49] It concluded that there was no need for a release of Wigner energy from the Calder Hall reactors for at least two years, but did not mention the civil Magnox reactors.[50] It emphasised the need for operational manuals both for normal running and 'infrequent maintenance work'. Mostly it dealt with Windscale Pile No.2 and the necessary conditions for restarting and operating it. The greater part of the text was taken up with explanations of Wigner phenomena and a discussion of different methods of annealing without nuclear heating, by blowing electrically heated air through the pile core.[51] Otherwise the text included little material from the working party reports, which the committee considered too technical and detailed for inclusion in full. This summary treatment left some of the conclusions and recommendations, especially on filters and cartridges, with little explanation but it avoided appearing to cast doubt on the White Paper of November 1957.

The White Paper of July 1958 advised that steps should be taken to anneal Pile No.2 by a method using heated air, but only after improved instrumentation and certain other modifications had been installed. It recommended[52] a substantial increase in the number of instruments (suggesting a total of 600 thermocouples) and said apparatus should be available at all times for detecting fuel element failures. It repeated that the filters had played an important part in reducing the hazard to the public from the Pile No.1 fire but recommended[53] that the existing filters should be improved; however, it did not refer to the working party's suggestion of a chemical scrubber system, (see above). It asserted that the uranium fuel cartridge design was satisfactory but that the current design of lithium-magnesium cartridge was unsatisfactory, and added that a full study must be made of the performance under all conceivable pile operating conditions of all materials before they were inserted in the reactor.

The Authority accepted the TEC recommendations.[54] At the press conference held to mark the publication of this last Fleck report, an Authority spokesman forecast the abandonment of Pile No.1.[55] Its rehabilitation had been very seriously considered but the difficulties of decontaminating the undamaged parts of the graphite core would have been immense, and the cost almost as much as that of building a completely new pile. The idea of using the pile stack as a radioactive waste silo was discussed, but rejected because in the long term it might be necessary to dismantle the stack. The Authority decided[56] not to try to restart Pile No.1 and announced the decision on 15 July. On 16 July *The Times* reported 'the end of No.1 reactor at Windscale – completed in 1950 at a cost of £3 700 000 and severely damaged in the accident last October'. As much technical information as possible was being got from the reactor core; useful material and equipment would be recovered; some of the buildings would be adapted to other uses; the rest would be sealed off. The future of the undamaged Pile No.2 was under review and the Authority estimated that it would need time – 2–4 months – to reach a decision.

THE IPCS DISAGREES

Before considering the fate of Pile No.2 we now turn to the IPCS and its activities. As Chapter 6 showed, the Windscale staff and the IPCS (their staff association) were far from happy about the first White Paper. Between 8 November and 21 February the Institution – at national and local level – carried on a voluminous correspondence, and had several meetings, with senior Authority representatives and, later, Authority Members.[57] A week after the White Paper appeared in November 1957 the IPCS sent the Authority chairman a memorandum,[58] 'Notes on Cmnd 302', which questioned it on nine points. IPCS objectives were to defend its members whose professional integrity had been impugned and who felt 'deeply wronged'; to point out the mistakes they perceived in Cmnd 302; and, they hoped, to make a constructive contribution to the Fleck Committee. Uncertain whether the White Paper accurately reflected the Inquiry findings, they asked to see the full Penney report, in order to study the basis for the summary version and for the opinions expressed in the White Paper, but the request was refused. They were at a loss to understand why the Authority's technical staff could not see information relating to their own plant and operations.

After a fruitless meeting with Sir Donald Perrott on 27 November[59] the IPCS wrote to the Prime Minister[60] asking him to receive a small deputation to put their request for a copy of the Penney report. No. 10 courteously referred the IPCS back to the Authority, saying that the latter would be arranging to hear complaints by the staff concerned, and would give them access to the technical records used by the Inquiry.

After further correspondence, sometimes touchy, Cockcroft, Perrott, Peirson and Mitchell (the IG secretary) met seven IPCS members at Windscale on 21 February 1958. A joint statement[61] issued afterwards said that in official reports on the accident there was no intention to reflect on the behaviour of staff; nevertheless 'the Authority maintained its view that errors of judgement were committed in the application of the second nuclear heating'. The Windscale staff were completely dispirited by this meeting, but the Authority representatives had suggested that some points of disagreement were suitable for reference to the Fleck Committee. So the IPCS submitted a detailed paper to the TEC and asked to give oral evidence.[62]

Their paper contained sections on the applied research effort, lack of consultation, errors in Cmnd 302, and conditions for the start-up of Pile No.2. A great deal of research work done by the Authority during and after the Penney Inquiry had, it said, produced new knowledge that could have been available earlier if more effort had been devoted to it. As it was, Wigner procedures had been built up pragmatically by experience on the piles themselves. But any operational peculiarity ought to be immediately investigated by competent research scientists so that Operations Branch could be provided with the most authoritative knowledge and advice; R & DB needed expansion to cover trouble-shooting as well as forward development. Consultation between operational and research staffs had been hindered by staff shortage which left the operators too little time for it.

The detailed IPCS criticisms of Cmnd 302 were of its description of a Wigner anneal as a controlled release, a routine maintenance operation, and self-sustaining; and of its comments on instrumentation, the second nuclear heating, the early failure of uranium cartridges, and the inoperability of the scanning gear.

A Wigner anneal, the IPCS argued, could not be called a controlled release; releasing stored energy from the graphite was a deliberate action but was controlled only in the sense that a maximum temperature was not to be exceeded. The rate of heating was a function of the pile itself. Neither was a Wigner release a routine maintenance operation, since each anneal produced its own characteristics and conditions had to be judged by the pile manager and consultant physicist. The release might be self-sustaining in a small body of graphite, but by no means so in a mass of 2000 tons.

As for instrumentation, the IPCS thought that having only one thermocouple on continuous recording was probably a factor in the accident; all thermocouples should be of the continuously recording type so that temperature changes in the pile would be immediately visible. The number and distribution of thermocouples was not necessarily inadequate; they had worked satisfactorily for many anneals. Two years earlier it had even been proposed to reduce the number because of the shortage of thermocouples, but this idea had been rejected. (Nevertheless, as we saw in Chapter 4, Gausden had recently asked for more.)

The main objection made by the IPCS was to the White Paper statement that the second nuclear heating had been applied too soon and too quickly and had led to the fire. It will be remembered that the second nuclear heating had been decided upon because it was essential to keep the Wigner release going, and the operators believed it to be dying away. The White Paper had said that a second nuclear heating was not appropriate at the time since a substantial number of graphite thermocouples were showing steady increases, and the general tendency was for graphite temperatures to be rising, not dropping. The IPCS paper disagreed. The 'substantial number', it said, was only $12\frac{1}{2}$ per cent of the total, and the remaining thermocouples were falling, steady or 'fluctuating around steady'. On previous experience, the operators concluded that in these circumstances more heating was necessary. The two officers who made the decision had had much experience of Wigner anneals and one had been present at all the anneals on both piles.

If not put in too soon, was the nuclear heating put in too quickly? Running up the pile was done at a set speed on a prescribed drill, the IPCS argued; the rise of temperature was a function of the pile itself, and what the operator did was to respond to the rise in temperature shown by the recording instruments. The operators had done this in accordance with their instructions.

The IPCS did not share the White Paper opinion that one or more uranium cartridges had failed on Tuesday, 8 October. The readings of counters placed underneath the stack filter to record radioactivity showed no rise until the dampers were opened for the fourth time on Thursday, 10 October (the fourth occasion when there was a positive air flow through the pile). If the cartridges had been burning for 36 hours some radioactivity would have been detected earlier by the counters, yet there was no record of it.

The final point of disagreement was on the BCDG. The jamming of the gear was not something unusual, as the White Paper implied, but an inevitable result of the high temperatures reached in the piles during anneals. It had occurred during every anneal, and after each one the gear had remained out of operation for some time.

In general, the IPCS said, Cmnd 302 was too categorical about the cause of the accident. On the facts presented, the cause was a matter of probability rather than certainty, and the evidence given in the White Paper did not support such emphatic conclusions.

Finally, the IPCS paper recommended that if and when Pile No.2 was restarted the pile manager should be given a guarantee that all the cartridges he had to put into the pile had been tested satisfactorily under all conditions of pile operation. It was not clear that this had been done in the past with lithium-magnesium cartridges.

The secretary of the TEC replied[63] to the IPCS head office about the Institution's memorandum. Two parts of the memorandum – on applied research and on consultation – could, he wrote, be dealt with by the strengthened headquarters recommended by the first Fleck report, and the remaining points would be passed to the TEC working parties. A few days earlier, writing to the IPCS branch at Windscale, Perrott had reaffirmed[64] the Authority's absolute adherence to the technical conclusions in Cmnd 302; in particular to the conclusion that errors of judgement in the application of the second nuclear heating were among the causes of the accident. IPCS hopes that these disagreements would be resolved by the TEC were finally dashed when the TEC secretary wrote[65] to IPCS head office on 20 March (incidentally confusing the Penney report and the first White Paper):

> It would not be appropriate for this committee to indicate whether anyone acted with outstanding merit or whether anyone committed an error of judgement . . . Its functions will

be fulfilled by its examination of all technical facts and records and . . . the submission of recommendations to enable the proper technical operation of reactors to be carried out in future . . . It is not some kind of Appeal Court or Revising Committee from the Penney Committee and in particular is not required to comment on particular actions or procedures *in the past* save in so far as they are relevant for technical action in the future.

PILE NO.2

The Authority studied the Fleck TEC recommendations on conditions for restarting Pile No.2 and the technical and economic problems involved.[66] The cost of providing the recommended equipment was estimated at £500 000 but, even as the Fleck report was being published, results were appearing that raised doubts about the future of the pile as a production unit. Would its prospective working life justify spending £500 000?

Harwell and the IG proceeded to carry out an intensive examination of the new information being obtained on the state of the graphite in Pile No.2. This pile, it may be remembered, contained purer graphite than Pile No.1, and had been in operation for a shorter time so that the graphite had undergone less irradiation.

Besides work on Wigner energy problems a study was made of the oxidation of graphite samples taken from Pile No.2. In eight channels examined, three graphite blocks were found for which the oxidation rates were some 3000 times greater than average; and probably the worst blocks had not yet been found; neither had the extent of the highly reactive regions. At such rates a piece of graphite would oxidise completely in 35 hours at 400°C. The graphite runaway oxidation temperature was estimated to be 380°C (if the block was burning at the surface only) or 320°C (if the block was burning throughout). However, in a triggered Wigner release the air temperature could not be expected to be much below 250°C. Therefore, between the minimum graphite temperature needed for a Wigner release and the runaway oxidation temperature, the difference might be no more than 50°C (or even less, since the samples examined so far did not represent the worst graphite in the reactor). On this latest evidence there was a strong risk of uncontrolled oxidation if graphite temperatures were allowed to exceed 300°C in a Wigner release. To reduce the risk to reasonable proportions the maximum temperature in a Wigner release should not be allowed to exceed 250°C. But then the stored energy that failed to be released would accumulate so that the life of the reactor would be too short to be worthwhile.

The only safe way to operate Pile No.2 in future as a production unit would require a preliminary anneal (the sixth for this pile) at 300°C, in the manner recommended by the Fleck TEC, using electrically heated air, *but after discharging all the fuel*. Even this procedure would not provide a guarantee against the very real risk of a fire due to runaway oxidation – and the destruction of the pile – but removal of the uranium fuel would prevent a radioactive hazard to the district in this event. If the pile survived this preliminary anneal it could be reloaded with fresh fuel and restarted, but would then have to be operated so that the graphite temperature was never allowed, accidentally or deliberately, to exceed 250°C. Since Wigner energy would build up again, low temperature anneals would be required (probably about every six months). If used not as a production unit but as a low power irradiation facility the pile would not need these frequent low temperature anneals; but the initial anneal, with its attendant fire hazard, would still be necessary. Moreover, the value of the irradiation facility would be very limited. In view of the risks, the cost, the operating difficulties and the uncertain future of the pile, it is hardly surprising that the Authority Board accepted the IG's recommendation not to restart Pile No.2. It was written off at £1.7 million, less such sums as could be recovered by the use of certain buildings and equipment (some £$\frac{1}{4}$–$\frac{1}{2}$ million). We shall return to these Pile No.2 investigations in discussing the possible causes of the Pile No.1 accident in Chapter 8.

In a speculative article on 16 October, the *Daily Telegraph* anticipated the closing of Pile No.2:

> Almost certainly it will be sealed and its fate will be the same as the No.1 reactor . . . The modifications recommended by the Fleck Committee to improve No.2 in the light of the accident would have cost the Authority a minimum of £500 000 . . . Any plutonium produced after all modifications had been made

would be more expensive than any other plutonium produced elsewhere in Britain. Initially Windscale was the only plant in Britain capable of producing the material for an atomic bomb. The four Calder Hall reactors would be producing it more cheaply and so would the four at Chapelcross when they come into operation. Also by 1962 the first plutonium would be coming, if the Government wished it, from the first of the civil power station reactors . . . Sir Christopher Hinton, chairman of the CEGB, said that No.1 would be a 'monument to our ignorance'. No.2 will now be another monument to keep it company.

The article went on to explain that Windscale was expanding and there was no likelihood of redundancies there.

Plowden submitted the Authority decision on Pile No.2 to the Prime Minister[67] on 17 October, the same day that planning permission for the new experimental advanced gas-cooled reactor at Windscale was announced. On 22 October the Prime Minister approved a press statement that Windscale Pile No.2 was to be written off as a total loss, and its demise was formally announced on 24 October. However, the work of the third Fleck Committee and its working parties was not wasted. It was followed up within the Authority and was usefully applied to the reactors at Calder Hall and Chapelcross and to the civil nuclear power stations. It led to many modifications in design and operation and especially in instrumentation, at an estimated capital cost to the Authority of £5.5 million over four years.[68]

The accident itself, and the four White Papers, focused Government attention and concern on problems in great need of them. They greatly influenced the Authority, strengthened its organisation, and quickly achieved the staffing improvements that Hinton and Owen had vainly sought for years, and they created a new awareness, in the Authority and to a lesser extent outside, of the importance of nuclear health and safety; consequently the development of the nuclear industry in Britain was undoubtedly safer than it would otherwise have been. But this was not all gain (see Chapter 10).

8 Causes: An Accident Waiting to Happen

What was the true cause of the accident? Given the original Government decisions to produce plutonium for atomic bombs and to build and operate air-cooled graphite moderated piles for the purpose, there are several possible and cumulative causes. There were failures of knowledge and research, especially on graphite. The piles were used for purposes not envisaged when they were designed and lacked adequate instrumentation. There were deficiencies in staffing, organisation, management and communications. The whole project was overloaded with too many urgent and competing demands - to expand and extend military production, develop new reactor types, and support an arguably over-ambitious civil power programme. This overloading bore most heavily on the IG and especially on Windscale, its most inherently vulnerable site.

All these more or less remote causes have appeared in earlier chapters. The immediate technical cause is the concern of this chapter, which briefly considers the evidence and various explanations put forward. However, it does not identify a single and certain cause. It seems clear that we still cannot say exactly what started the 1957 fire; various scientists hold differing views on the most probable cause: some favouring graphite, some a lithium-magnesium cartridge, some a uranium fuel element. From among the mass of data that accumulated in the year after the accident, evidence can be found to support each of these. We may never know the answer, even when decommissioning of the pile is completed. Whatever the initiating cause, the sequence of events in the pile core must have been highly complex and interactive.

The first and only published statements on the subject are contained (as we saw in Chapter 6) in the original White Paper (Cmnd 302). The immediate cause of the accident, it said, was:

> the application too soon and at too rapid a rate of a second nuclear heating to release the Wigner energy from the gra-

phite, thus causing the failure of one or more cartridges in the pile, whose contents then oxidised slowly, eventually leading to fire in the reactor ... The accident was due partly to inadequacies in the instrumentation provided ... and partly to faults of judgement by the operating staff, these faults of judgement being themselves attributable to weaknesses of organisation.[1]

The White Paper discussed the possible cause or causes in four other passages (my emphases):

> The *immediate cause of the accident was the second nuclear heating*. This was applied before it was necessary and the nuclear heating was put in at too rapid a rate ... The second nuclear heating led to the accident by causing the failure of one or more cartridges whose contents then oxidised slowly, eventually leading to the fire. *By far the most likely possibility is that the second nuclear heating caused the failure of one or more uranium fuel cartridges.* A second possibility, which cannot be eliminated, is that it was a lithium-magnesium cartridge which failed. The second nuclear heating may have released pockets of Wigner energy at a time when the general level of temperatures throughout the pile was high.[2]

> *As a result of the second nuclear heating* [Tuesday, 8 October], *graphite temperatures gradually rose through Wednesday 9 October. This led to the oxidation of the uranium which had been exposed by the overheating*. The exposed uranium smouldered throughout the course of the day, 9 October, and gradually led to the failure of other cartridges and to their combustion, and to the combustion of the graphite. By Thursday evening [10 October] the fire had spread and was affecting about 150 channels.[3]

> *Whilst the argument is one of probability rather than certainty*, ... the most likely cause of the accident was the combined effect of the rapid heating and the high temperature reached by the fuel elements in the lower front part of the pile. In all probability, one or more end caps of the cans of fuel elements were pushed off, and uranium exposed ... The cause of the accident could conceivably have been *the failure of a lithium-magnesium cartridge* but ... this was *unlikely*.[4]

The possibility that, irrespective of the second nuclear heating, a *large local release of Wigner energy* occurred in a pocket of graphite that had not been annealed for some time, and that as a result high local temperatures were caused, [*was rejected*].[5]

Thus the White Paper emphatically identified as the probable immediate cause the second nuclear heating on the morning of Tuesday, 8 October. This operation, it judged, was both premature and incorrectly carried out, owing to errors of judgement on the part of the operating staff, and it had caused a uranium fuel cartridge – or more than one – to fail and to smoulder undetected, until the fire was discovered two days later, on Thursday, 10 October.

These conclusions in the White Paper were based on – but went rather beyond – the report of the committee of inquiry. The Penney Committee,[6] conscious of public anxiety about the accident, had felt obliged to report as soon as it had considered the technical evidence sufficiently to discharge its terms of reference. However, in less than ten days it could not, it said, make 'a full technical assessment of all aspects of this matter', and it recommended that a technical evaluation committee within the Authority should carry out 'an urgent and thorough study of all the technical information to be derived from the accident'. This task was later remitted to the Fleck Committee (see Chapters 6 and 7).

Even before the Penney Inquiry was set up, as we saw in Chapter 5, the IG began its own technical studies of the accident, which were essential for its continuing and future operations. In addition, a crash programme of graphite research – crucially important to the civil nuclear power programme as well as the Authority's reactors – was put in hand. Most of this experimental work was absorbed by the specialist working parties serving Owen's internal technical executive committee (TX) and later Sir Alexander Fleck's committee (TEC), and the accident studies were given an over-riding priority.

Although the objectives of these various investigations were not identical with those of the Penney Inquiry they often had to deal with similar technical questions, but they had the advantage of more time than the Penney Inquiry. They were also able to call on specialists in all the relevant fields, whereas the inquiry had had limited resources and no specialist working parties. In early December 1957 valuable information was gained from detailed exchanges of information with American experts who came to England to discuss the accident (see Chapter 6). Later, too, the results of many more experiments and calculations became available, and important new data came from graphite samples and irradiated cartridges taken from the shut-down Pile No.2.

It would have been remarkable if so much effort, for many months after the accident, had failed to throw any new light on it. Professor Diamond – a member of both the Penney Inquiry and the Fleck Committee – told the latter that he had:

> stressed the importance of the Authority's undertaking the most thorough examination of all the information that had been made available to the Penney inquiry in order that the maximum information could be extracted from the accident ... The Penney inquiry had had to examine and comment on this information at high speed. A less hurried study of all the records would undoubtedly provide valuable information ... relevant to the points made by the IPCS ... [and] of great benefit for the safe operation of reactors in future.[7]

SOME SCEPTICS

From the beginning not everyone accepted all the White Paper conclusions. Gethin Davey, the works general manager, did not; D. R. R. Fair, formerly manager of the Windscale Pile Group and since May 1957 at Chapelcross, did not; and J. C. C. Stewart, Director of Technical Policy at Risley, did not. Davey was convinced the fire was caused by the failure of an AM cartridge. Fair, in his copy of the White Paper,[8] pencilled marginal comments. Against the statement that it was unlikely that a failed AM cartridge had been responsible, he wrote: 'If this unlikely then conditions for failure of U cans more unlikely.' Against the paragraph rejecting the idea of a large local release of Wigner energy he simply wrote '20/53', referring to the thermocouple which had shown abnormally high temperature readings during the anneal (above 420°C for $3\frac{1}{2}$ hours, and above 400°C for 17 hours). In only one previous anneal (Pile

No.1 in November 1956) had the maximum graphite temperature exceeded 400°C; it had then recorded 400°C for seven hours and 420°C for not more than a quarter of an hour.

J. C. C. Stewart, by no means convinced that the conclusion reached by the board of inquiry was correct,[9] proposed that a team led by his deputy, Dr (now Sir John) Hill, should go over the available evidence. This was done and Hill's appraisal (see below) was presented to Owen on 17 January 1958.

The IPCS, as we saw in Chapter 7, alleged mistakes of a technical or factual nature in the White Paper account, and thought its categorical statements about the cause of the fire should have been at best statements of probability. The IPCS did not point to any single cause, but thought it inconceivable that if uranium cartridges had failed on 8 October they could have smouldered undetected until 10 October. It also observed that, in dust samples taken from the cyclone in the pile stack, lithium was found in significant quantities before any increase in fission products. This fact strongly suggested that, at the time of sampling, AM cartridges had already caught fire; but the evidence was inconclusive because comprehensive air samples were not available.[10]

REINTERPRETING THE EVIDENCE

The Bowen Memorandum

Meanwhile at the end of November J. C. Bowen, of the recently formed Safety Branch at Risley, reinterpreted the evidence and offered a more complex scenario.[11] His memorandum suggested that after the second nuclear heating on the morning of 8 October some graphite temperatures had increased by about 100°C over a period of 32 hours, because of the combined effects of fission product heat and self-heating of graphite caused by oxidation; the latter would be insignificant at 300°C but important at 370°C. Then, at about 2 a.m. on 10 October, in a region of the pile where the ambient temperature was about 400°C, a fuel cartridge burst. By 7 a.m. half of one channel was glowing completely, the graphite temperature had risen by 60°C and the adjacent AM channel by 30°C. The general temperature level rose and at 460°C some AM cartridges ignited.

The fire then spread outwards, igniting more lithium-magnesium (AM) cartridges. When the temperature reached 650°C it was hot enough to melt the aluminium cans of the uranium fuel elements. When a high flow of coolant air was permanently switched on, at 2 p.m., it cooled down some regions of graphite, where the temperature was below 450°C. But graphite above this temperature – that is, all the graphite in the vicinity of the burning lithium-magnesium – continued to burn, and so did the uranium fuel.

This admittedly speculative account agreed with the first White Paper in pointing to a uranium cartridge as the origin of the process, but attached more importance to the AM in spreading the fire, and left open the question of the second nuclear heating.

The Hill Memorandum

After completing the re-examination undertaken for Sir Leonard Owen, J. M. Hill presented 'A more detailed assessment of the early stages of the Windscale accident'[12] on 17 January 1958. His most important conclusion, he said, was that no substantial oxidation of uranium fuel occurred before 5 a.m. on 10 October, when an increase in activity on the stack filter chambers was recorded. Oxidation of uranium had taken place fairly late in the chain of events. He did not see how a few smouldering uranium cartridges could readily have led to the uniform temperature that was actually observed in the core, and he did not believe that there had been a uranium fire undetected in the core for more than a very short period. In his opinion the BCDG, even though jammed (as always during Wigner anneals), could still have detected a serious burst.

The AM cartridges were, he believed, a principal factor. Their integrity was very dependent on temperature; they could stand 425°C for a very long time but would quickly fire in experimental rigs at 450°C (a much lower temperature than that at which uranium fuel cartridges would normally fail). The AM cartridges could not, however, have been the sole cause of the fire. Hill concluded that a fire in a single AM channel had led to oxidation of the graphite immediately surrounding it, and this had provided most of the heat necessary to propagate the fire. Probably much of the lithium-magnesium had burnt, and there had been sufficient slow graphite oxidation to raise the temperature by a few tens of degrees Celsius, before serious oxidation occurred in the uranium, resulting in the main fire.

From the point of view of the disaster, Hill said, it did not matter very much what was the order of conflagration of the lithium-magnesium, graphite and uranium. He thought his conclusions did not conflict with those of the Penney Inquiry, 'which attributed the accident to a failure in a cartridge, which could have been either an AM cartridge or a uranium cartridge'. Nevertheless, his account of the probable chain of events differed markedly from that in the first White Paper.

Did Hill's appraisal have any bearing on the second nuclear heating, or the alleged mistakes by the operating staff? It was clear, he said, that the rapid rise in temperature during the second nuclear heating was due to the abnormal manner in which the pile was being started up, with the pile operator deliberately trying to produce a distorted flux pattern. But this was agreed procedure for Wigner releases, intended to secure a release of stored energy from regions of the pile where it had especially accumulated. The pile had come to criticality more rapidly than had been expected; the low power instrumentation (boron-10 counters) normally used during routine – that is, non-Wigner – start-ups was not in position since it would not withstand the temperatures encountered in the pile during anneals. The over-swing in temperature, Hill said, could have been avoided by the use of the shutdown rods, but this was not done because it would have seriously delayed the second nuclear heating; in any case the over-swing, as recorded on the instruments available to the operator at the time, did not appear serious. The most important handicap for the operator had been insufficient evidence on the exact temperatures in the pile (rather than the jamming of the BCDG). In spite of the 70 or so thermocouples installed there were large areas of the pile where the temperature was most uncertain, and the area of the fire was in fact in a region between thermocouples.

The Hill memorandum implied that because the instrumentation provided for use during Wigner anneals was so inadequate, the operators had acted reasonably on the information available to them at the time. If so, they could not be accused of faults of judgement. Whatever went wrong the men operating the pile had in fact complied with their written instructions and had used their considerable combined experience of previous Wigner anneals (difficult and unpredictable operations that were far from the controlled releases or routine maintenance suggested by the White Paper).

The TEC Cartridge Working Party

The next study relevant to the cause of the accident was by the TEC's expert working party on cartridges.[13] It ruled out isotope cans (other than AM) as a hazard. In uranium fuel cartridges, it said, failures during normal operation were due less to temperature than to irradiation growth; below a certain irradiation level the burst rate appeared independent of pile operating temperatures. The design of the uranium fuel element was satisfactory, but catastrophic oxidation could occur even at low temperatures, and therefore safety must depend on detecting and discharging an oxidising element before it ignited other cartridges or the graphite moderator. A maximum fuel temperature of 395°C, on past experience, was safe in normal operations but potentially unsafe in shutdown conditions with the BCDG out of action.

In AM cartridges lithium/aluminium compatibility was a serious problem. Post-accident experiments had found some diffusion of lithium from the rod into the aluminium can at temperatures as low as 250°C; above 437°C the can might be penetrated in less than an hour, exposing the alloy. Temperatures over 400°C had been recorded in some anneals and so the Mark III AM cartridge (in which the alloy and the aluminium can were in direct contact) was dangerous unless pile operations were severely restricted. A safer design was essential.

Mark III AM cartridges had gone without apparent trouble through a Wigner release in Pile No.2 in July 1957, when the graphite temperature had not exceeded 305°C. Later, however, when AM cartridges were discharged from Pile No.2 after it was shut down, some were found to be severely damaged. This information was too late for the Penney committee to take into account and it might have modified its conclusions as a result, as Penney commented when he received photographs of the cartridges. However, there was no sign of the damaged cartridges having caused graphite oxidation.

The cartridge working party, like the TEC itself, was concerned only with the safety of future operations, and did not express an opinion on the cause of the 1957 accident. Its report implied that either a failed uranium fuel element or a failed AM cartridge might have initiated the fire, but found the design of the latter intrinsically dangerous.

Another Explanation?

People continued to hold differing opinions, but the question of what caused the fire in Pile No.1 was effectively closed. However, another possible explanation is that the fire originated in the graphite itself and was not caused initially by the failure of a cartridge, either uranium or AM. This would account for the absence, until a late stage, of radioactivity readings on the stack counters or static monitoring apparatus. There are other arguments which also support this explanation.

The Penney Inquiry had rejected the idea of an initial graphite fire, but had lacked important information which only became available much later. Weeks after the TEC report, and nearly a year after the Penney report, IG and Harwell scientists completed a thorough investigation of Pile No.2, which was still shut down.[14] Their object was to examine the feasibility of annealing the pile by the method the TEC recommended – using heated air blown through the pile, instead of nuclear heating – with a view to restarting and running it under the operating conditions prescribed by the TEC. These conditions included radically improved instrumentation, redesigned BCDG and hundreds of additional thermocouples (see Chapter 7).

Samples of graphite had always been routinely taken from both piles for examination, but with Pile No.2 shut down for a long time much more extensive sampling was possible and a great deal was learnt about the condition of the graphite. Even so the scientists emphasised that their knowledge both of Wigner energy storage and of graphite oxidation was still very limited.

As we saw in Chapter 7, they found the graphite in Pile No.2 to be so affected by years of irradiation that its resistance to oxidation was greatly reduced, and they concluded that in this condition the margin between the temperature needed to ensure a satisfactory Wigner release and the temperature liable to cause runaway oxidation was dangerously narrow. Even if all the uranium fuel was discharged before annealing the pile with heated air, there would still be an unacceptable fire risk.

Pile No.2 had been running for six years and four months when it was shut down in mid-October 1957; the graphite in Pile No.1, after seven years' operation, had probably been in worse condition when the unlucky ninth anneal began. But investigations of the graphite in Pile No.2 which led to its permanent closure in October 1958 were directed towards future operations and, apparently, their relevance to the cause of the 1957 accident – by then a closed question – was never considered. However, supposing the margin of safety had been as narrow then in Pile No.1 as it was later found to be in No.2, the operators of the older pile had, unknowingly, been in a Catch-22 situation. They could not carry out a Wigner release at all without risking a graphite fire, but to run the pile without a Wigner release also meant risking a graphite fire. If there was real danger of a fire in Pile No.2 even if it was annealed with all the uranium fuel removed, the odds had certainly been against annealing Pile No.1 safely by nuclear heating. From the viewpoint of October 1958, allegations of faults of judgement by the operating staff look thoroughly unconvincing. Whatever the operators had done or not done on 8 October 1957, a fire was surely inevitable; if not then, soon afterwards. The 1957 Windscale fire really was an accident waiting to happen.

CONCLUSIONS

Arguments about the cause are still inconclusive and some of the evidence is conflicting. Some points to AM cartridges; for instance, the detection of lithium in the pile stack cyclone before any fission products, investigations reported by the cartridge working party, and the state of the AM cartridges recovered later from Pile No.2. Davey, Stewart, Hill and Fair all attached special importance, in varying degrees, to AM cartridges. The Penney Committee, however, had thought it much more likely that a uranium fuel element initiated the fire. It had also ruled out graphite oxidation, but the later evidence from Pile No.2 supported graphite oxidation and this opinion is strongly held by at least one physicist who was there at the time; normally graphite is very difficult to burn but, as we saw, the graphite in the Windscale piles was not normal after several years of operation.[15]

During the entire period from the end of the second nuclear heating on Tuesday, 8 October, to the discovery of the fire on Thursday, 10 October, Hill wrote, heat was clearly being produced from the burning of AM cartridges, the oxidation of graphite and the combustion of fuel elements, and the whole reactor was very finely balanced between safety and catching fire.[16] The order of events contributing to the combustion of lithium-magnesium alloy, graphite and uranium was not important, except for its bearing on the recommissioning of Pile No.2. But, as we know, Pile No.2 was never restarted because even the conditions recommended by the Fleck TEC report were too stringent to be practicable and further operation would have been uneconomic.

Once the Authority had decided not to restart Pile No.2, did it matter for practical purposes that the exact initiating cause of the accident was still uncertain? Probably not, as far as the Magnox reactors were concerned. Higher inlet operating temperatures were expected to reduce or even eliminate the storage of Wigner energy in the graphite and hence the need for anneals; in the event this expectation was amply fulfilled and no anneals were ever required. But the reactors differed from the old Windscale piles in important respects, as we saw in Chapter 6. The use as a coolant of carbon dioxide in a closed circuit, instead of air discharged to the atmosphere, reduced or almost eliminated the risks both of graphite oxidation and of radioactive emissions. The accelerated research programme prompted by the Windscale accident also provided valuable and timely information for application to the Authority's own reactors at Calder Hall and Chapelcross, and to the civil power station programme. The dangerous Mark III AM cartridges were discontinued. Instrumentation was greatly improved. Millions of pounds were spent on modifications to enhance the safety of the Authority and CEGB reactors.

In all the circumstances, the question of the immediate cause of the 1957 fire may seem to be of academic interest only. It made sense for the Fleck TEC report to disregard it if it was irrelevant to future reactor operations. Such scientific speculations would not have interested the public, which was concerned simply to know whether or not nuclear reactors, especially the civil power stations, were safe. The public had, said Plowden, been peppered with statements about Windscale, and now the last thing anyone can have wanted was to reopen questions 'settled' by the first White Paper in November 1957.

The only people who would have welcomed a reopening of these questions were the IPCS and some Windscale staff. They firmly believed that a re-examination of the cause or causes of the accident would vindicate them. Later evidence does so, but Cmnd 302 remains the last published word on the subject. Does this matter? Again, perhaps not for practical purposes, but after more than 30 years it is surely time to do justice to the Windscale men. They had worked immensely hard to meet all the demands of the nuclear defence programme, which for post-war British governments – Labour and Conservative alike – was a matter of the highest national importance and priority. Furthermore, after an accident which had become inevitable, they had acted with outstanding courage, resourcefulness and devotion to duty. Yet their actions had been publicly blamed, at the highest level, as contributing materially to the fire. Perhaps the present account will put events into a new perspective.

9 Appraisals and Reappraisals

The Windscale piles have been closed down for over 30 years and questions about the exact cause of the accident may seem academic; however, its health impact is still very much a live issue, and several assessments of it – from zero upwards – have been made (see Appendix X). Variations over the years are due, as we shall see, to three factors: recognition of pathways to human beings previously discounted; revised estimates of the quantity and composition of the radioactive emissions from the damaged reactor; and – most importantly – changing assumptions about the relationship between dose and effect at low levels of radiation.

The first assessment of health consequences was made on 15 October 1957, only three days after the fire was quenched; the second was made four days later; there were two more before the end of 1957; the most recent was in 1988. They differ in purpose and method. The earliest appraisals were made quickly as a basis for practical counter-measures; later studies were done to gain a better understanding of the event or to assist in planning for any future emergency.

Some appraisals were based on survey results – the measured concentrations of the more important radioactive materials on the ground, in the air, or in milk, water, herbage and certain foodstuffs – and on human radiation exposures, measured *in vivo*. Later assessments also used estimates of the total radioactivity released to the environment. Although the early work was concerned with individual radiation exposures, later studies emphasised the collective exposure of the population.

Scientists carrying out the later studies had more time and more information at their disposal as well as new techniques and computing facilities not available in 1957. The radiological context, too, was changing and new methodologies, concepts and standards were evolving. Direct comparisons are therefore more difficult to make than might appear at first, especially as the units of measurement have been altered in the meantime.[1]

FIRST THOUGHTS, 1957

The original health impact assessment was made by the Authority's health and safety staff and four of its radiological consultants[2] (see Chapter 5). Meeting at Risley on Tuesday, 15 October, they first discussed the analyses of Friday evening milk samples from Seascale farms, which showed a concentration of 0.4 microcurie of iodine-131 per litre. On Saturday and Sunday, radioactivity had been expected to rise; it had done so by a factor of four or five, and by Tuesday was at a plateau. Local milk supplies, as we saw, had been banned during the weekend; the milk survey was extended as rapidly as possible, and further restrictions were imposed on distribution of milk containing more than 0.1 microcurie of iodine-131 per litre. Andrew McLean, the IG's chief medical officer, explained to the consultants how he and his colleagues had calculated this limit in order not to exceed a dose of 10 rads to an infant's thyroid. A later account by Dunster and colleagues (see below) gives 20 rads[3] (see Appendix VIII). A calculation at the time by the late Sir Edward Pochin suggested that the 0.1 microcurie concentration would produce an average integrated dose of 7 rads; this compared with a probable carcinogenic dose of 250 rads.

The consultants agreed that the 0.1 microcurie action level was reasonable. No other foods appeared to present a hazard, and radioactive substances other than iodine-131, they decided, were comparatively unimportant. Young children and babies were most vulnerable because of their relatively large milk consumption and the small size of their thyroids. The milk ban based on 0.1 microcurie of iodine-131 per litre had apparently kept the dose to infant thyroids down to 20 rads or less, a level considered innocuous, so it could be safely assumed that the public had suffered no harm.

The consultants' meeting on 15 October 1957 was followed on 19 October by another meeting of experts,[4] convened at the request of the Ministry of Agriculture and chaired by Sir Harold Himsworth, the Secretary of the MRC (see Chapter 5). They were satisfied, on the information available, that there had been no danger to public health, but recommended for future action a limit of 0.06 microcurie of iodine-131 per litre of milk instead of 0.1 microcurie.

The Penney report, presented to the Authority on 28 October (see Chapter 6) concluded that there had been '*no immediate danger to the health of any of the public or of the workers at Windscale*' and that it was '*most unlikely that any ill-effects [would] develop*'(my emphasis.)[5] Ministers then sought the formal advice of the MRC, and the November 1957 White Paper contained the views of an MRC Committee, again chaired by Sir Harold Himsworth. Summarising its findings it said, '*The information available is adequate to allow an assessment to be made of the possible risks to human health and safety arising from the recent accident at Windscale. After examining the various possibilities we are satisfied that it is in the highest degree unlikely that any harm has been done to the health of anybody, whether a worker in the Windscale plant or a member of the general public* (my emphasis).[6]

REPORT TO THE GENEVA CONFERENCE, 1958

Among the British papers presented at the second international conference on Peaceful Uses of Atomic Energy in Geneva in the summer of 1958 was one by Dunster, Howells and Templeton on the Windscale accident and the subsequent district surveys.[7] Some of the information it contained had already appeared in the 1957 White Paper, but the Geneva paper explained in more detail how the Authority staff had calculated their emergency level for iodine-131 in milk, and also provided new information about the radioactive release. The fission products listed were iodine-131, caesium-137, strontium-89 and strontium-90, with negligible quantities of ruthenium-103, ruthenium-106, zirconium-95, niobium-95 and cerium-144. Also listed was polonium-210 which was not a fission product, but a radioisotope manufactured by irradiating bismuth oxide cartridges in the pile. The magnitude of the radioactive release from the pile stack during the accident was estimated as follows (in curies), with no figures given for polonium-210 or the 'negligible' fission products:

Iodine-131	20 000
Caesium-137	600
Strontium-89	80
Strontium-90	9

These figures were not presented as having any bearing on the health effects of the accident; from the health point of view the criterion was the level of individual exposures. For those plainly at risk (local children and some Windscale employees) these had not exceeded a level considered harmless by the best medical opinion. Thanks to the preponderance of short-lived iodine-131 in the release and to the milk ban, few children seemed likely to have received more than 10 rads. The Windscale laboratories had made 238 thyroid measurements of adults and children living downwind of the plant; the highest dose to a child's thyroid was about 16 rads, and to an adult's 9.6 rads. Most were much lower. A systematic follow-up of these 238 people would probably have been regarded at the time as both unnecessary – since, according to the medical authorities, no harm was to be expected – and undesirably alarming, to the general public as well as to the families concerned.

Confidence seemed solidly based on the concept of a 'permissible dose' of ionising radiation, which 'in the light of present knowledge [was] not expected to cause appreciable bodily injury to a person at any time during his life-time'. Appreciable bodily injury was defined by the ICRP as any effect 'that a person would regard as being objectionable and/or competent medical authorities would regard as being deleterious to the health and well-being of the individual'.[8] Clearly any effects that might conceivably result from radiation exposure at, or below, the permissible dose were thought to be very minor; most certainly not such effects as thyroid cancer or leukaemia.

FURTHER THOUGHTS FROM THE MRC, 1959–60

In December 1960 the MRC published the second report[9] by its Committee on the Hazards to Man of Nuclear and Allied Radiations, which was chaired by Sir Harold Himsworth, the first report having appeared in June 1956.[10]

The 1960 Himsworth report included (as an appendix) a paper[11] previously published in the *British Medical Journal* in April 1959. It was a report by an expert panel set up by the MRC's Committee on Protection against Ionising Radiations (PIRC) to advise on emergency levels of dietary contamination in the event of a nuclear reactor accident. The recommended limit for iodine-131 in milk was 0.06 microcurie per litre, the same as the MRC limit proposed in October 1957; it could, however, be relaxed provided children under 3 years old were fed on dried milk or fresh milk from an uncontaminated source.

The Himsworth report also contained (Appendix H) a nine-page paper on the environmental aspects of the Windscale accident.[12] It updated the 1958 Geneva estimates of the total radioactive release; the figures for iodine-131, caesium-137 and strontium-89 were unchanged, and for strontium-90 the figure was reduced from 9 to 2Ci. However, a new component, tellurium-132 (12 000Ci), appeared for the first time. Again no estimate was given for polonium-210, which was not mentioned. It has been suggested[13] that there was a cover-up of polonium-210 (because it was a secret atomic weapon material, or even because the British did not want the Americans to know that they were still using such an old-fashioned ingredient in atomic bombs). This suggestion seems implausible because the 1958 Geneva Conference paper by Dunster, Howells and Templeton referred to polonium, and the polonium-210 release had also been mentioned in a paper by two Harwell scientists, Stewart and Crooks,[14] published in *Nature* in 1958; but it is possible that these references would not have appeared if it had not been known that some Dutch scientists had detected polonium-210 in the atmosphere just after the accident and were publishing the details.[15]

A probable explanation of the omission of polonium-210 is that, in the over-riding concern about hazardous fission products – especially iodine-131 and strontium-90 – polonium-210 was not thought important. Two Harwell scientists, Marley and Stewart, certainly discussed it in 1957 and Marley, Harwell's chief radiological safety officer, advised Cockcroft that compared with iodine-131 it need not be considered a significant hazard in milk.[16] Nevertheless it is incomprehensible that it should have been omitted in 1960 in what was obviously intended as a definitive account. The omission had later repercussions.

The 1960 MRC report reaffirmed that the only danger from which the public had had to be protected was milk contaminated with iodine-131. This, it said, had been successfully controlled by the milk ban. In a few instances, where there had been reason to suspect that the milk control might not have been fully effective, the people concerned – and other representative members of the public living downwind – had been invited to the Windscale laboratories for thyroid measurements, with the results already reported at Geneva.

The 1960 report reiterated the conclusions in the 1957 White Paper that 'the information [was] adequate to allow an assessment to be made of the possible risks to human health and safety arising from the ... Windscale accident', and that it was *'in the highest degree unlikely that any harm [had] been done to anybody, whether a worker in the Windscale plant or a member of the general public'* (my emphasis).

The report also revealed that the 1957 accident was not the only source of radioactive contamination. Some was due to world-wide fallout from Russian and American nuclear weapon tests; strontium-90 found on farms ten miles or more from Windscale was mainly from this source. But the highest contamination by strontium-90 (confined to a few fields in the immediate neighbourhood of the piles) was mainly caused by the hitherto undisclosed emissions of fairly coarse particles of nuclear fuel from the pile stacks in the two or three years before the accident (see Chapter 3). At times the milk from farms within two miles of Windscale had up to ten times more strontium-90 than the milk from comparable farms elsewhere in the United Kingdom. However, the report considered the average level of strontium-90 in milk to be well within the limits then admissible.

If the revised (and generally higher) emergency levels recommended by PIRC in Appendix J of the 1960 MRC report are applied retrospectively to the October 1957 fire, only iodine-131 had approached the permissible limits. It was estimated that the mean radiation dose to the thyroid glands of a few children reached half the maximum acceptable dose of 25 rads; as mentioned above, the *highest* measured dose to a child thyroid reported by Dunster and his colleagues in the Geneva paper was 16 rads.

Appendix H of the 1960 report, on the environmental aspects of the accident, did not mention collective dose but this new concept, and the no-threshold hypothesis from which it derived, were discussed elsewhere in the report. Up to 1958 the agreed definition of permissible dose (or maximum permissible dose) implied a radiation threshold below which no harmful effect – or 'appreciable bodily injury' – was expected. Then, in its 1958–59 recommendations,[17] the ICRP redefined the permissible dose as *'that dose, accumulated over a long period of time or resulting from a single exposure, which, in the light of present knowledge, carries a negligible probability of severe somatic or genetic injuries ... Any severe somatic injuries (eg leukaemia) that might result from the exposure of individuals to the permissible dose would be limited to an extremely small fraction of the group'*; permissible doses were *'expected to produce effects that could be detectable only by statistical methods applied to large groups'* (my emphases).

The 1960 Himsworth report concluded that the term threshold, and the idea of minimum radiation doses below which no harmful effect would be produced, could no longer be sustained. Though sufficient knowledge existed to assess, approximately, the damage from high and moderate rates of exposure, the likely degree of damage from low rates of exposure could only be inferred. (This is still true in 1991.) It was therefore thought prudent to assume a linear dose–effect relationship, with even the lowest radiation exposures involving a correspondingly low risk of injury in the long term. Thus the size of the dose was related to the statistical probability, and not the severity, of adverse effects.

This linear hypothesis – generally thought to be conservative and thus to over-estimate rather than under-estimate risks – had important implications. As long as there was believed to be a threshold – provided doses were below the limit – radiation workers were considered to be safe for a working life-time. *A fortiori* the public, receiving much smaller doses than radiation workers, would be protected and – even if quite large numbers of people were exposed to low levels of radiation – no damage to health would result as long as individual doses did not exceed the limit.

Without a threshold this view was no longer tenable; even a small exposure carried a risk, however infinitesimal, of serious long-term harm. The number of people exposed and the total radiation received had become important, not simply individual doses. If there was *some* risk even at very low doses, then the more people received such doses the greater the chance of someone suffering injury. The concepts of collective dose and of risk coefficient (or risk factor) followed logically. Inevitably this new thinking would sooner or later (though not in the 1960 MRC report) be applied to the Windscale accident. It would

Appraisals and Reappraisals   143

increase the significance of the release itself (its magnitude and composition) compared with all the other data, including individual exposures. New assessments of the release were made from 1963 onwards, but no new appraisal of its health impact appeared until 1981. In the meantime a great deal more radiobiological knowledge had accumulated, especially about internal radiation and pathways to human beings.

REASSESSMENTS, 1963–76

In 1963 J. R. Beattie, a Risley scientist, produced a reactor safety study[18] which included incidentally a re-examination of the fission product releases during the Windscale accident as estimated by Dunster and his colleagues in 1958. Like them, Beattie based his work mainly on district surveys of deposited radioactive material and on sampling of stack filters (see Appendix IX).

He also calculated the total quantity of fission products in the damaged fuel, the proportions retained by the filters, and the amounts released to the atmosphere. Much the largest components in the fuel, he wrote, were strontium-90 (380 000Ci) and cerium-144 (290 000Ci), but of these only 0.02 per cent and 0.03 per cent reached the atmosphere, whereas 12 per cent of the iodine-131 present did so. His calculation of the amounts of iodine-131, tellurium-132, caesium-137, strontium-89 and strontium-90 released to the atmosphere was the same as that given in the 1960 MRC report, but he added ruthenium-106 (80Ci) and cerium-144 (80Ci).

Beattie did not mention polonium-210. His purpose was to assess environmental hazards from fission product releases; polonium-210 was not a fission product and he was concerned with the safety of current and future reactors (in which polonium-210 was not produced) and not with a *post mortem* on the defunct Windscale piles.

An important new assessment of the Windscale release was made eleven years later by Roger Clarke (then a CEGB scientist and subsequently Director of the NRPB).[19] Using all the previous data, with further data based on the fuel inventory and the temperatures recorded during the fire, his computer calculations largely confirmed the 1958 and 1963 assessments. Significantly, Clarke then argued the importance of the inhalation pathway, previously considered negligible compared with ingestion. He emphasised, too, the implications of small radiation doses to large numbers of people – the collective dose – and of organs other than the thyroid. Overall, he thought the doses to lung and large intestine might have been more significant than thyroid doses.

The collective dose and the inhalation pathway were again emphasised by Baverstock and Vennart (of the MRC's Radiobiology Research Unit) in a 1976 report,[20] although they had insufficient data to quantify the risks:

> If, as is presently agreed by most responsible authorities it is assumed . . . that effects might be linearly related to dose, *the collective dose (person-rem) received by persons remote from the site of an accidental release might be of greater concern than that received by those nearer to the site and subject to the counter-measures.* Doses of up to 0.1 rem might have been received in the thyroids of millions of people in Lancashire, Yorkshire and Southern England, and doses about an order of magnitude less could have been experienced in Belgium. *The total effect of these numerous very small doses would have been several fold greater than the effect of the larger doses of up to 10 rem in relatively small populations near to Windscale.* (My emphases)

AFTER THREE MILE ISLAND, 1979–88

Another major reactor accident occurred in March 1979 in the United States. At the Three Mile Island power station, at Harrisburg, Pennsylvania, a reactor came near to a core meltdown. It was an entirely different type from the old Windscale piles. A pressurised water reactor (PWR) of an up-to-date design, it was fuelled with enriched uranium, moderated and cooled by light (ordinary) water with the coolant in a pressure circuit, and enclosed in a strong containment structure. The amount of iodine-131 released to the atmosphere during the accident was very slight: just 16Ci compared to Windscale's 20 000. Nevertheless it was a major reactor accident and could have been a disaster if the hydrogen bubble above the reactor core, in the primary containment, had exploded.

At the same time (and 24 years after his original involvement in work on the Windscale release) a Harwell scientist, Arthur

Chamberlain, redirected his attention to some of the questions that still puzzled him. In particular, why had the 1957 radioactive release not been even larger than it was, given the fission product inventory of the many channels affected by the fire? He studied the release again,[21] returning to all the contemporary evidence available: from grass and soil samples; from the radiological survey above the pile stack, made by helicopter within days of the fire; from samples taken from the stack filters; and from an air filter at Calder Hall. He also had the results of many other air samples taken at the time: some at Windscale for routine health physics purposes, some for monitoring fallout from nuclear weapon tests, some taken by local authorities to monitor smoke pollution, and some collected from various European towns.

Chamberlain concluded that for the most part substances that were volatile at 1000°C had gone up the pile stack; hence the great preponderance of iodine-131 in the release, the small proportion of strontium and the virtual absence of uranium oxides and plutonium. Of the total iodine-131 in the 140 affected channels – some 112 000Ci – he calculated very approximately that 27 000Ci (25 per cent) had been released to the atmosphere; 16 000 to 17 000Ci (15 per cent) had been washed out of the pile and discharged to the sea; 64 000Ci (57 per cent) had been retained within the pile; and 4000Ci (4 per cent) had been retained by the filters (see Appendix IX). The pile filters did no more than stop a fraction of the radioactivity that reached them (something the Penney Committee could not have known).

The stack filters certainly made some contribution to safety during the accident, and it was as well they were there. However, they played nothing like the crucial part that had been thought and is still widely believed. Indeed, since they had signally failed to stop particulate emissions that had persisted for over two years before the accident, they could hardly be expected – even with the improvements then in hand – to meet the stringent test of the fire; and, as the Fleck filter working party made clear, no filters could have done.

After the Three Mile Island incident a major official reassessment of both the Windscale release and its health impact was undertaken by the NRPB. At the same time an unofficial study was put in hand in 1979, when the Union of Concerned Scientists in Massachusetts commissioned a new study of the 1957

Windscale accident by Peter Taylor of the Political Ecology Research Group (PERG) in Oxford. Taylor found that no computation of the collective dose and no numerical assessment of the health impact had ever been published, and he proceeded to fill this gap, and was indeed the first to try to do so. For reasons of chronology the PERG report,[22] published in July 1981, is here dealt with before the NRPB reports of 1982 and 1983.

Using all the existing scientific literature, including the meteorological data, Taylor computed collective doses of the many radionuclides released; he referred to polonium-210 but concluded that it was of no particular significance. He then applied the current (1977) ICRP risk coefficients both for whole-body radiation exposure and for exposure of individual organs. On this basis he calculated that the radiation exposures received by the United Kingdom population from the Windscale fire could result in between 10 and 20 deaths, occurring over several decades. The number of thyroid cancer cases might be up to 250, of which 10 might be fatal; this is a rare disease, usually non-fatal, and as it has a long latent period any cases would only become apparent in the decade after 1977. There might also be one fatal lung cancer as a result of the Windscale release. Such figures, Taylor said, would not be detectable in national or regional health statistics.

If the current ICRP risk factors were too low (as many scientists then thought), and if thyroid cancer mortality was 10 per cent – instead of 5 per cent – then the upper bound could be 280 deaths over several decades. This excess would still be invisible against an *annual* cancer mortality of some 140 000 in the United Kingdom. In short, even on the most pessimistic assumptions, Taylor did not expect the Windscale fire to have any detectable health impact. Reviewing national and local statistics he found no significant deviation from the expected rates in the areas affected by fallout from the fire.

The NRPB study[23] by Crick and Linsley was published in 1982. It emphasised the significance of the collective dose in assessing health consequences, while pointing out that after a nuclear accident counter-measures would always, and rightly, be aimed at reducing individual risks. It investigated the saving in radiation exposure achieved by the milk ban and calculated that *the ban had halved the collective dose in West Cumbria and had reduced the total collective dose by 12 per cent.*

The two NRPB authors gave considerable weight to the inhalation pathway, to external radiation, and to organs other than the thyroid. They estimated the upper bound of thyroid cancers in the United Kingdom at 260, thirteen of them fatal (compared with an *annual* incidence of 650 a year in England and Wales). The upper bound for all fatal cancers they calculated at 20. (Unlike the PERG report, they did not produce figures for Western Europe.)

The NRPB report did not mention polonium-210 and after articles in the press drew attention to the omission – alleging a cover-up – the two authors produced an addendum to their report in the following year.[24] This 1983 addendum took account of polonium-210, tritium and very small amounts of plutonium and uranium. Though iodine-131 was still the main contributor (37 per cent of the total dose), it was now followed closely by polonium-210 (36 per cent). The recalculations led to an increase of 67 per cent in the estimated collective dose,[25] and thus to higher estimates of the health consequences. The upper bound of fatal cancers rose from 20 to 33, an increase that was still within the margin of accuracy claimed for NRPB's 1982 figure. To put the polonium-210 release into perspective: over a lifetime it would result in a radiation dose one-tenth of the *annual* dose received from naturally occurring polonium-210, which on average contributes one-tenth of the background radiation from *all* natural sources.

The NRPB was to make later reassessments of the Windscale accident in 1988–89; meanwhile Windscale was coming under scrutiny from other directions.

WINDSCALE UNDER SCRUTINY

The general health consequences of Windscale/Sellafield have been much in the news for years. But this book is about the 1957 Windscale accident, not about wider questions of environmental health, and it must pass over recent reports and press articles except where they are relevant. In November 1983 a television programme, 'Windscale – the Nuclear Laundry', presented details of young people in the area who had been diagnosed before the age of 22 as suffering from various types of cancer, including leukaemia, between 1954 and 1983. The programme argued that, in the absence of any other obvious cause, a link with radioactive discharges from Windscale ought to be seriously examined. (The PERG report had drawn attention to leukaemia clusters, which Taylor thought needed study, although he did not believe they were connected with Windscale or the 1957 accident.) The government responded immediately and the Minister of Health appointed an eminent physician, Sir Douglas Black, to chair an independent advisory group. It reported in 1984 on the possible increased incidence of cancer in West Cumbria.[26]

The Black report took account of the 1957 Windscale fire, among all the other sources of radiation to which West Cumbrians born between 1945 and 1975 had been exposed. These other radiation sources included routine discharges from the Sellafield site, fallout from nuclear weapon tests, medical radiation and natural background. Background radiation was by far the largest source. The contribution of the 1957 fire, Black found, was generally much lower than that from any other sources although it accounted for 53.9 per cent of the total thyroid dose, and 5.3 per cent of the total lung dose. The 1957 fire contributed little to organ doses likely to cause leukaemia.

The Black report recommended further epidemiological studies of the local population, increased efforts to improve radiation dose assessment, and the creation of a national body to deal with the adequacy of data and the need for more research. The Committee on Medical Aspects of Radiation in the Environment (COMARE) was quickly set up. Its first report,[27] also published in 1984, did not examine the 1957 fire separately but studied the total effect of releases from the plant. COMARE had some new data, discovered too late for the Black advisory group to use, concerning pre-accident emissions of particles from the pile stacks (see Chapter 3). From information provided by a former Windscale scientist, Dr D. Jakeman,[28] who had worked on stack filter problems, it appeared that 20 kg of uranium oxide had been emitted from the pile stacks between 1954 and 1956, rather than the 440g estimated in 1957. More data were then found, and analysed for COMARE by Chamberlain,[29] when quantities of old papers were being reviewed in preparation for transfer to the PRO.

COMARE was disturbed by the inadequacies and uncertainties of the available data, which compounded the intrinsic difficulties of assessing health effects. The committee observed with truth that, in making dose and risk assessments, '*a very complex chain of reasoning, involving many uncertainties, is necessary to go from the release data and the sparse environmental monitoring data to a prediction of any possible adverse health effects*' (my emphasis).

Nevertheless the committee felt able to conclude that the increased radiation doses during the 30-year period of 1950–80 (including those due to the fire) were still well below those the population received in the same period from natural radiation and nuclear weapon test fallout combined.

As a consequence of the rudimentary monitoring methods of the 1950s and the inadequacies of record keeping, COMARE said, 'we shall never know the actual average doses received by the population', and 'it is likely that we shall never be able to establish with any certainty whether there is any relationship between the cases of leukaemia in Seascale and the radioactive discharges from the Sellafield site'. However, information about the 1957 fire, and estimates of its health impact, are probably more complete and accurate than any other environmental data for Windscale's early years.

A study of childhood leukaemia in the neighbourhood of nuclear establishments was published in January 1988 by three NRPB authors.[30] They considered the 1957 Windscale fire among the sources of man-made and natural radiation to which children born and living in Seascale were exposed between 1945 and 1979. They calculated the *total* risk of radiation-induced leukaemia in this population as 1 in 12 250. This figure is lower than the incidence actually observed but, whichever figure is taken, what is significant for the present purpose is the contribution of the different sources, estimated as:

Natural sources	approximately 66 per cent
All Sellafield/Windscale discharges (including the 1957 fire)	16 per cent
Weapon tests fallout	9 per cent
Medical procedures	9 per cent

Of the 16 per cent from all Windscale/Sellafield discharges, some 1.8 per cent is attributable to the Windscale fire.

Thus, the radioactive release to the environment due to the fire made virtually no contribution to the risk of childhood leukaemia in Seascale. This is what one would expect since the release contained mainly thyroid-seeking iodine-131 and only very small quantities of bone-seeking (and therefore leukaemogenic) radionuclides.

The puzzling discrepancy between the observed and the expected excess incidence of childhood leukaemia, referred to by the NRPB authors' January 1988 report, may be linked with the results of a new epidemiological study[31] undertaken by Professor Martin Gardner and colleagues in direct consequence of the Black report. The Gardner report finds the raised incidence in West Cumbria to be associated not with radioactive discharges and environmental pathways but rather with the fathers' employment at Sellafield and the external doses of whole body radiation recorded before the children's conception. Further work is clearly required to identify if the observed association is indeed due to external radiation, or to intakes of radionuclides such as plutonium; whether it was possible that contamination reached the developing embryo/foetus; finally, whether it may have been due to chemical agents to which the workers were exposed reaching the sperm or foetus. The chemical possibility especially needs to be explored since Gardner also notes excesses of leukaemic births to workers in farming and in the iron and steel industries.

The external radiation doses of Windscale workers were routinely measured by film badge and were recorded for 13-week control periods, for which the limit, recommended by ICRP, was 3 rem; anyone exceeding the 3 rem limit was removed from contact with radiation for a compensatory length of time. For the whole of the control period during which the accident happened, fourteen men involved in it received over 3 rem, the highest recorded dose being 4.66 rem, and the usual practice of temporary suspension from radiation work was followed so that the accident should not have added to the cumulative external radiation exposures of individual Windscale workers. Mortality studies of 470 employees involved in the Windscale fire show that cancer deaths to October 1987 were below the 'expected' number.[32]

THE WORST NUCLEAR ACCIDENT IN THE WORLD, 1986

The world's third major reactor accident[33] – and immeasurably the worst – occurred at Chernobyl, in the Ukraine, on 26 April 1986, in a 1000-megawatt water-cooled, graphite-moderated reactor, a system which, it will be remembered, the British had rejected on safety grounds in 1947 (see Chapter 2). An experimental operation at low power led to a sudden explosive power surge. Radioactive materials were thrown high into the atmosphere. Unlike Three Mile Island, the Chernobyl reactor had no outer containment building. The graphite moderator, which normally ran very hot, caught fire. It burned for several days, until cooling could again be established, while radioactive emissions from the core continued, peaking again on 4–5 May.

Thirty-one men on the site died and two others died later; about 300 people suffered from radiation sickness. Some 150 000 had to be evacuated from their homes within a 30km radius of the power station, an immense logistical task. It is expected in the Soviet Union that the health of more than 600 000 people who may have incurred significant radiation will be monitored throughout their lives. Latest calculations of health consequences among the Chernobyl evacuees suggest some 700 additional cancer deaths over the next 40 years or so, compared with a natural expected incidence of some 27 000. Nearly 3 million acres of agricultural land is lost for decades because of contamination with radioactive caesium, strontium and plutonium. The true picture of the Chernobyl radioactive fallout in the Soviet Union is emerging only now, several years after the accident.

The total fission product release was estimated at 40 000 000–50 000 000Ci, well over 1000 times that from Windscale. Radioactive contamination was severe; it was also very widely dispersed, causing enhanced radiation levels in many countries as the radioactive clouds blew across Europe. Due to wind direction and the incidence of rainfall there were marked discrepancies between, and within, countries; there were also marked discrepancies in collective doses, depending on population distribution, lifestyle and diet. Within the United Kingdom, for instance, the rain-enhanced deposition was greater in the north and north-west than in the south and east.

For the United Kingdom the upper limit for possible cancer deaths from Chernobyl fallout, based on collective dose, is estimated at about 40 (using the ICRP 1977 risk factors) rising to 100 (using new and higher risk factors). Either figure is statistically invisible against a background of some 142 000 natural cancer deaths annually (or 5½ million in 40 years).

FINAL WORDS ON THE HEALTH IMPACT?

Changes since October 1957 in assessing the possible health impact of the Windscale accident have resulted, as we have seen, partly from revised estimates of the release itself (its magnitude and composition) and from consideration of new pathways to human beings. The inclusion of polonium-210 and the lung cancer risk exemplify both.

The principal changes have, however, resulted from developing ideas about low-level radiation and dose–effect relationships. A major shift in thinking, as we saw, was the abandonment by about 1960 of the threshold hypothesis that dominated official thinking in the 1950s, and the progressive adoption of the concept of collective dose. This meant that it was no longer possible to say that the health impact of the 1957 Windscale accident was nil.

However, what the dose–effect relationships were was a matter for much argument. ICRP had produced a wide range of risk coefficients but, as data accumulated, especially from the life-span epidemiological study of the Japanese atom bomb survivors, some scientists believed that the ICRP risk coefficients were too low and thus led to an under-estimate of the health impact of any given collective dose. In 1987 the NRPB recommended,[34] in advance of any ICRP revision, that risk coefficients should be raised by a factor of two or three, and in 1988 UNSCEAR[35] revised its own coefficients.

In October 1988 the director of NRPB published a new paper[36] on the health consequences of the 1957 accident using the 1988 UNSCEAR risk coefficients instead of the ICRP risk coefficients used hitherto. From the collective dose equivalent, Clarke derived new figures for the upper bound of cancer deaths in the United Kingdom (100 over a period of 40–50 years) and non-fatal effects (90 non-fatal cancers and 10 hereditary defects).

Of the 90 non-fatal, 60 per cent would be thyroid cancers (say, one or two a year over 40 years, compared with a natural incidence of 650–700 a year). Of the fatal cancers the most significant number would be lung cancer from polonium-210, but it would mean less than one case a year in the United Kingdom, compared with a total fatal lung cancer rate of about 41 000 a year. Clarke commented: '*At the very low levels of individual dose involved, it must be questionable as to whether the risks are real. They can only be regarded as an upper estimate and the most likely number of health effects will be lower and may be zero* (my emphasis)'.

To recapitulate: it seems unlikely that any health effects will ever be observed which can be attributed to the Windscale accident. This is not to minimise the Windscale fire. It was a serious accident, but its consequences could have been very much worse. If the main component of the release had not been short-lived iodine-131; if the stack filters had not been in place (minor as their contribution was); if the milk ban had not been effectively imposed; above all, if the Windscale men had failed to control the fire before it got out of hand, then the accident would have been an environmental catastrophe. It undoubtedly yielded invaluable and timely lessons on nuclear safety, just as Britain embarked on an ambitious civil nuclear power programme. It was extremely fortunate that it did this with so little human damage, or perhaps none.

10 Postscript

THE END OF AN ERA

Thirty-five or so years after the Windscale accident its exact initiating cause remains uncertain, as appeared from Chapter 8. More, and conclusive, evidence may be found when the planned dismantling of the two piles is completed. The accident was almost certainly inevitable. It was undoubtedly salutary, and its ill-effects were very much less than they might have been. It was, fortuitously, timely. It marked the end of an era.

The accident was by hindsight inevitable, or virtually so, in consequence of the over-loading of organisation and staff, and insufficient knowledge and understanding of complex phenomena. But the over-loading was not uniform; it especially affected the IG's operational staff at Risley and factories.

How was this allowed to happen? The Authority knew it (see Chapter 3). Hinton and Owen certainly knew it, but knew too that meeting the requirements of the defence programme was considered paramount. 'Shortages of men, materials or knowledge', Owen wrote, 'were not allowed to jeopardise it. The times given were such that risks had to be accepted.'[1] The piles were under particular pressure as they were the bottleneck in the whole defence programme; all the other plants had spare capacity. On these two piles depended the weapon test programme, the H-bomb development, the bomb stockpile and the V-bomber force currently being created; in short the entire British nuclear deterrent and Britain's position *vis-à-vis* the United States as well as the Soviet Union.

This inexorable defence programme was only part of the Authority's, and the IG's, ever-growing workload. Up to 1953 (apart from some of Harwell's research activities), the atomic energy project had had a single-minded purpose: to produce a British atomic bomb. From 1954, when the AEA was set up, the overall programme became more and more diverse as well as larger. To an expanding and increasingly complex and demanding military programme were added an ambitious civil power programme, and the development of new and more advanced reactor systems, besides all Harwell's varied research projects and the opening up of new sites. In retrospect it is surprising that all this was undertaken when the Authority's manpower was known to be below strength, when recruitment was frustratingly difficult, and when the national shortage of qualified engineers and the weakness of the British engineering industry were only too apparent (see Chapter 3).

The Authority's manpower resources were inadequate for its total commitments, and priorities were not firmly applied within it. Hinton, looking back from his new vantage point at CEGB, wrote[2] to Fleck after the accident that the Authority's programme of R & D and engineering work had been allowed to grow too quickly, and to a size not justified by the volume of business that would arise from it. If Authority resources had been concentrated more rigorously on the defence programme and the industrial power programme, he thought, manpower shortages might have been manageable. More staff would have been available for operations, and the load on the R & D and engineering staffs would have been reduced. But with new sites opening at Dounreay, Calder Hall, Chapelcross, and later at Winfrith, experienced men were continuously transferred from operational work to new projects (Windscale, it was said, was 'milked'); and top management, as well as research and safety staff, were too over-loaded with work and too occupied with new plant and future systems to think about existing plant. (When news of the accident reached Owen at Dounreay, he is remembered as saying that he had not thought about the Windscale piles for years.)

The first Fleck report (Cmnd 338) was emphatic on priorities. The most important responsibility of the IG was to run the large-scale plants already in operation efficiently and safely. This was the Authority's, as well as the Group's, corporate responsibility. If these large-scale plants and current operations had been given their proper priority – in high quality staff, research effort and equipment (such as pile instruments) – the Windscale accident might conceivably have been avoided.

The follow-up to the accident was turned to good account: so much so that, technically, it could be – and was – described as the best of reactor experiments, albeit an involuntary one. Besides the technical information acquired, the important research initiated and the organisational reforms, it created a heightened consciousness of health and safety and a new 'safety culture' which did not exist before the Windscale accident, however conscientious and careful individual staff might be. In a potentially hazardous industry this 'culture' requires safety to be given an over-riding priority. It demands well-qualified specialist staff in sufficient numbers, and strong and independent health and safety organisations able to withstand operational or other pressures and to apply a critical expertise to plant design, construction and operation. It must be based on mandatory and enforceable standards, in addition to self-regulation (even though self-regulation alone may be unavoidable for a time in a new technology).

After the 1957 accident these conditions began to be met. As we saw, the Authority created a central health and safety branch (the AHSB), independent of the Groups and individual sites, with a Radiological Protection Division and a Safeguards Division[3] – the forerunner of the Safety and Reliability Directorate – concerned with reactor and plant safety. The new AHSB quickly began work on a comprehensive series of codes of practice on radiological protection and the design, commissioning and operation of reactors, chemical plant and laboratories. It also had to ensure that the enormous gaps in the Authority's emergency planning were filled; it may be remembered that in October 1957 when the accident occurred, only Windscale, of all the Authority's sites, had an emergency plan.

On a national basis, the Nuclear Installations Inspectorate (NII) was created in 1959 (although not directly as a result of the accident).[4] It had comprehensive licensing and inspecting powers, which were undertaken by the Authority on an agency basis for three or four years until the Inspectorate itself was staffed and organised to carry out its duties (although until recently the Authority itself was formally outside its jurisdiction). From 1960 onwards statutory regulation began to replace self-regulation in the industry. Eventually, in 1971, the NRPB (itself a late offspring of the 1957 accident) came into being and absorbed most of the old RPD.

One of the most serious deficiencies revealed by the accident was in relevant research. This fact is not inconsistent with Hinton's view, noted above, that R & D in the Authority had grown excessively; a great deal of research was going on in the Authority, but too little of it was directed to practical needs or to the support of essential current operations. The 1957 accident demonstrated the importance to a high-technology industry of adequate and well-focused research and close co-ordination between R & D and operations. It stimulated a very considerable amount of applied research within the Authority which was of great value both to the Authority's reactors and to the Magnox reactors of the first civil nuclear power programme; but for the Windscale accident it might well not have been done. The accident itself, the TEC working parties and the TEC report led to important improvements in operating procedures and especially in instrumentation; reactors in Britain were all undoubtedly safer as a result of the accident. The lessons of Windscale 1957 were well learnt.

As for the radiological safety aspects, the accident stimulated work on the setting of appropriate safety standards, which as we saw were not available in October 1957. (This deficiency had not in fact led to any failure to protect public health as the *ad hoc* standard applied in imposing the milk ban was actually more rigorous than the MRC's 1975 limit: see Appendix VIII.) Other radionuclides than iodine-131, other pathways, and the new concept of collective dose, were taken into account in the later assessments of health effects. Even so, in his independent and unofficial study in 1981 Taylor – not an apologist for the nuclear industry – concluded that the maximum number of fatal cancers that might result from the Windscale accident over several decades would not be detectable in national or regional health statistics. In 1988 Clarke, the director of the NRPB, using new and more pessimistic risk coefficients, calculated possible totals of 100 fatal and 100 non-fatal effects over a period of 40–50 years, but he questioned whether, at such very low levels of individual dose, the risks were real; his figures, he said, represented an upper limit and the likely number of health effects would be lower (perhaps even zero).

In financial terms the cost of the accident was not large as the two piles which had to be written off were quite uneconomic by comparison with the Calder Hall reactors and would have been closed down fairly soon in any case. The main cost was for condemned milk, for which some £60 000 was paid, and a few small compensation claims (for example, for damage due to gates being left open, or to minor business losses).

The post-accident reorganisation was not entirely beneficial. The Fleck structure was perhaps cumbersome and overelaborate (see Appendix II). Moreover, although the Authority – and certainly the IG – had been seriously under-staffed, some people thought that the rapid expansion after 1957 went too far and left the Authority with major staff and organisation problems. Perhaps the Authority had not needed a massive increase in overall staff so much as a massive redeployment, against which its decentralised style and structure militated.

If the accident had happened later, the valuable lessons derived from all the consequent investigations might not have been in time to benefit the Calder Hall and Chapelcross reactors and the earliest Magnox power stations. For some of the latter the deadline for design modifications left little margin. However, if the accident had happened earlier, it would have dealt a crippling blow to the weapons development programme and so to the British nuclear deterrent. As it was, the supply of materials was not seriously interrupted by the loss of the two Windscale piles. By that time one Calder Hall reactor was in operation and the other three – with four more at Chapelcross – were soon to be in production. Aldermaston was able to conduct its tests. Then followed the negotiations with the United States leading to the crucial bilateral agreement on nuclear defence on July 1958. Without the tests which clearly demonstrated Britain's nuclear status, it is unlikely that this agreement – a prime objective of British governments since 1946 – would have been concluded.

'May I send you my congratulations?' Plowden wrote to the Prime Minister. 'It is a major achievement in the field of Anglo–American relations and one that has eluded successive British governments since the war.' 'I am grateful', Macmillan replied, 'to you and all those who helped you. You have done a fine job.' The IG, and Windscale, had contributed immensely to the fine job. The years 1946–57 had been a time of heroic effort.[5] With limited resources, formidable tasks in a new and unprecedentedly exacting technology had been undertaken as an urgent national duty, and had been brought to a successful conclusion by dedicated staff working under the most severe strain. It had also been a time of risk taking, as Owen said. Schonland commented that Davey was too used to getting away with risks,[6] but Davey was a careful man with much experience of running a wartime munitions factory and would not have taken avoidable risks. Britain was not alone in taking risks under inexorable atomic pressure.

It had also been an era of illusion when, to the makers of policy, all things seemed possible: that Britain should be a great military power with an independent nuclear deterrent; and simultaneously could gain a world lead, and an industrial advantage over the United States, by constructing the world's first civil nuclear power programme – twelve power stations in ten years – while forging ahead with more advanced reactor systems. The triumphant opening of Calder Hall in 1956 and the success of the weapon development programme both appeared to justify the confidence of the policy-makers. Sir Henry Tizard – scientific adviser to the Ministry of Defence – had warned that if Britain ignored her post-war limitations she risked the fate of Aesop's frog, who blew himself up and burst.[7] But the Authority scientists and engineers had met near-impossible demands and their very success in doing so invited yet further demands; Pelion was piled on Ossa. This era in British atomic history ended with the Windscale accident.

THE WINDSCALE PILES: PRESENT AND FUTURE

The familiar Windscale skyline is already changing, and soon the hawk-haunted chimney tops with their characteristic filter galleries will disappear. (Sizeable chimneys of a conventional shape will be left.) Since they were closed in 1957 the two old piles have remained inert, with their control rods fully run in and the drive mechanisms removed. They have been kept under constant surveillance, monitored and regularly inspected to ensure that they continue to be structurally sound, that radioactive contamination is not and cannot be released, and that there is no risk of gas explosions in the core, graphite oxidation or spontaneous Wigner releases. (Substantial amounts of Wigner energy are presumably stored in Pile No.2, and some in parts of Pile No.1.)

Since 1971 the responsibility has been shared between the Authority (which owns the pile buildings and the old blower houses) and BNFL (which inherited the chimneys and the pond), and the two bodies work closely together to ensure the safety of the piles. A recent inspection found that the brickwork and some structural steelwork at the top of the chimneys was deteriorating and it was decided to remove the top sections, including the filter galleries. Work has already begun on Pile No.2 stack, and part of the filter gallery has gone.

The piles are being completely isolated from the stacks, and special ventilation plants with absolute outlet filters are being installed before any work is begun which might disturb residual contamination. Then the pile cores (in which radioactivity has decreased to 1 per cent of the original level) can be examined in more detail. Only when more precise information is available on the state of the graphite in both piles (and of the fuel still left in Pile No.1) can a long-term strategy for decommissioning be formulated. The present programme of work will secure the long-term safety of the piles, whatever the strategy, and will supply the information necessary to define and implement further stages.

A special manipulator has been purchased to be used in removing undamaged loose fuel from the discharge face of Pile No.1, and modern remotely controlled underwater vehicles – originally developed for North Sea oil operations – have been used to explore the inaccessible water ducts and collect samples for analysis. Similar equipment will be used for cleaning up the ducts. The pond is being refurbished to enable the remaining fuel recovered from the piles to be safely transferred for reprocessing.

Options for the more distant future depend on many factors: the knowledge gained as a result of the present programme of work, which will extend over several years; the possible benefits of delaying decommissioning to permit further radioactive decay; and whether or not a suitable national radioactive waste repository is available then. Options will include maintaining the piles, under continuing surveillance, after the present programme of refurbishment is completed; or partial decommissioning, with the biological shields of both piles permanently sealed; or complete decommissioning, returning the pile area to a 'green field' state.[8] Whatever the final decision – and it need not and cannot be made for some years – safety, both of people and of the environment, must be the paramount concern.

Appendix I: Chronology of Events, October 1957– October 1958

1957

6 October	Preparations for ninth anneal of Windscale Pile No.1.
8 October	Second nuclear heating (11.05 to 19.25 hours).
10 October	Overheating in pile core. Radioactivity detected on site in air samples and on-site meteorological station. Health physics survey begun.
	Area of pile found to be on fire. Action taken to determine extent of fire, then to create fire-break. Attempt to discharge fuel elements from 120 affected channels.
	About midnight, decision to use water on fire if necessary. Just before midnight, Chief Constable of Cumberland warned of possible district emergency.
11 October	Evacuation plans ready, police on alert. Preparations to introduce water into pile.
	Water turned on 09.00 hours, and applied for 30 hours.
	Fire abated by midday. Chief Constable told emergency over. Report on accident sent from Windscale by IG Director of Operations to Chairman in London and Group Managing Director (at Dounreay).
	Prime Minister informed immediately. Press statement released.
	Windscale health physics manager begins biological monitoring survey, especially of milk. Risley health and safety organisation now involved. Milk samples to Harwell for analysis.
12 October	Water turned off. Clearing up operations begin.
	Results of milk analysis show unacceptable levels of iodine-131 in milk; *ad hoc* limit calculated for iodine-131 in milk. Arrangements put in hand to stop consumption of local milk: ban extends over 80 square miles.
	USAEC offers help and asks for information.
13 October	Press Conference in London.
14 October	AEA Chairman meets representatives of MAFF, MHLG

Appendix I

	and Ministry of Health to discuss health aspects of accident.
15 October	Milk ban extended to 200 square miles. Extensive programme of district surveys begun.
	Meeting at Risley of AEA's radiological consultants and AEA senior health and safety staff confirms milk ban and *ad hoc* limit of 0.1 microcurie iodine-131 per litre.
	Press informed that board of inquiry will be set up under chairmanship of Sir William Penney.
	IG sets up its own technical committee and plans special programme of research into technical aspects of accident.
17 October	AEX discusses the accident.
	Penney Inquiry begins work at Windscale.
18 October	Pile No.2 temporarily shut down.
19 October	Helicopter survey flight over Windscale pile stack.
	Meeting of MRC panel generally confirms views of consultants' meeting on 15 October.
23 October	In Washington, Eisenhower and Macmillan issue Declaration of Common Purpose on atomic co-operation for defence purposes.
26 October	Penney Inquiry completes report.
28 October	AEX considers Penney report and recommends publication in full. Report sent to Prime Minister.
29 October	Prime Minister has preliminary discussion of report with AEA chairman.
	Prime Minister asks Lord President of Council to obtain immediately views of MRC on accident and Penney report.
	Ministry of Defence agrees to publication of report.
	Penney report circulated widely to Ministers and senior officials; covering letter says publication is intended.
8 November	First White Paper (Cmnd 302) published.
10–11 December	USAEC scientists at Risley for US/UK Conference on accident.
17 December	First Fleck report (on AEA organisation) submitted to Prime Minister (published as Cmnd 338, December 1957).
20 December	Second Fleck report (on health and safety) submitted to Prime Minister (published as Cmnd 342, January 1958).
21 December	Interim report of Fleck TEC submitted to Prime Minister (not published).

1958

5 January	US scientists visit UK for conference on health and safety aspects of accident.
17 January	Reassessment of accident by Dr Hill submitted to Sir Leonard Owen.
February	French scientists invited to discussions on the accident.
March	Wigner release in BEPO at Harwell.
17 June	Report of Fleck TEC presented to Prime Minister.
15 July	TEC report published (Cmnd 471).
	Announcement that Pile No. 1 would not be rehabilitated.
August	Paper by Dunster, Howells and Templeton on district surveys following the Windscale accident presented at 1958 Geneva Conference on Peaceful Uses of Atomic Energy.
24 October	Announcement that Pile No.2 would not be restarted.

Appendix II: Responsibilities and Organisation of the IG

(The source of this Appendix is Appendix II of Cmnd 338 of December 1957: 'Report of the Committee appointed by the Prime Minister to examine the Organisation of certain parts of the United Kingdom Atomic Energy Authority'.)

Responsibilities

The responsibilities of the Industrial Group of the Atomic Energy Authority include the following items:

(1) The design, building and operation of defence installations;
(2) The design, building and operation of chemical plants to produce fuel elements for and to process irradiated fuel from the civil power stations;
(3) In conjunction with the Research Group, to design and build and, if necessary, operate advanced reactor experiments or prototypes;
(4) To collaborate with industry on the design and construction of nuclear power plant;
(5) To advise the United Kingdom electricity authorities on the nuclear aspects of the Power Programme and to advise overseas purchasers of reactors from United Kingdom industrial firms;
(6) To advise Government Departments and other organisations who may have an interest in the techniques of atomic energy.

The Production Executive Committee

The Production Executive Committee is the controlling committee of the Industrial Group. It is under the Chairmanship of the Managing Director and its members are the Executive Heads of individual branches. It works within the overall policy and financial control of the United Kingdom Atomic Energy Authority in London.

Within the framework mentioned above the Production Executive Committee formulates group policy in the above spheres, produces master programmes and co-ordinates the activities of the specialist branches of the group. The Production Executive Committee gives financial sanction for capital expenditure in aid of its programmes up to its delegated limit. It produces the Industrial Group annual budget for presentation to the Authority and controls both capital and operating expenditure within the Group.

ORGANISATIONAL CHARTS

On the following pages will be found charts showing:

(A) Present Organisation of Senior Staff at Industrial Group Headquarters.
(B) Proposed Headquarters Organisation of the Industrial Group.
*(C) Proposed Organisation of Operations Branch at Headquarters.
*(D) Proposed Organisation at the Works.

*Not reproduced here.

CHART A

PRESENT ORGANISATION OF SENIOR STAFF AT INDUSTRIAL GROUP HEADQUARTERS

Managing Director
- Director of Engineering
- Director of Research and Development
- Director of Technical Policy
- Deputy Director Industrial Power
- Director of Operations
- Director of Health and Safety
- Director of Administration
- Director of Accounts and Stores

Appendix II

CHART B

PROPOSED HEADQUARTERS ORGANISATION OF THE INDUSTRIAL GROUP

All those shown would be members of the Board of Management of the Group.

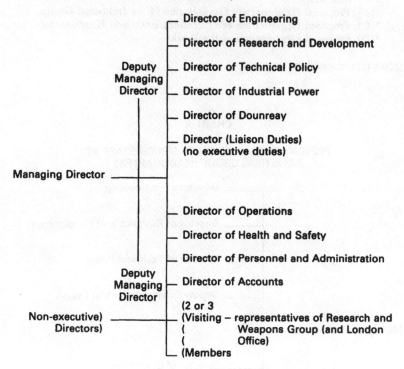

Appendix III: Instruction of 14 November 1955 on Wigner Releases

Mr Gausden

Wigner Energy Release

Will you please issue the following operating instructions to the operator engaged in controlling the Wigner Energy Release. If the highest Uranium or Graphite temperature reaches 360°C, then Mr Fair, Mr Gausden and Mr Robertson are to be informed at once, and the PCE alerted, to be ready to insert plugs and close the chimney base. When the maximum temperature reaches 380°C unless further instructions to the contrary have been received the roof plugs will be inserted and the chimney base closed. At 400°C all of the dampers in the blower houses are to be opened and at 415°C four shutdown fans are to be started up.

D R R Fair
Manager, Pile.

Copy to Mr Robertson

DRRF/RMcC   *November 14th, 1955*

Appendix IV: Summary of Wigner Releases in Windscale Piles, 1953–57

Table A.1 Summary of Wigner releases

		Maximum uranium temperature recorded	Maximum graphite temperature recorded	Time before release occurred (hrs)	Time for release to spread (hrs)	Time graphite remained hot (hrs)
Pile No.1						
18 August	1953*	360°	320°	?	?	
1 October	1953*	307°	339°	?	?	
6 July	1954†	295° (318°)	354° (334°)	5 (4)	21 (8)	24
14 March	1955	360°	330°	4	17	27
14 November	1955	295°	305°	1	16	21
6 February	1956	395°	340°	34	20	42
9 April	1956	370°	325°	–	–	–
30 November	1956	370°	420°	3	12	42
7 October	1957					
Pile No.2						
9 January	1953*					
25 November	1953*		320°			
24 August	1954	251°	310°	3	10	–
2 May	1955	300°	335°	5	24	–
23 January	1956	285°	375°	2	16	34
24 September	1956‡	390°	315°	4	27	29
1 July	1957	385°	305°	10	28	23

* I have found no detailed summary of these early releases. There had been an experimental release on 10 January 1953 in Pile No.1.
† The July 1954 anneal was in two halves: first the top of the pile, then the bottom. The pile was blown cold in between. The figures in brackets refer to the second half of the anneal.
‡ Second nuclear heating was necessary.

Appendix V: Note on Uranium Fuel Cartridges in the Windscale Piles

FUEL ELEMENTS

1. During the lifetime of the two Windscale piles, $\frac{3}{4}$ million fuel elements were used. They were all made originally of rods of *natural uranium* metal sealed in aluminium cans – finned to assist heat transfer – with a thin graphite lining to prevent chemical reactions between the aluminium and the uranium. Various marks were designed to improve performance, and altogether ten different marks were designed, and six were used, between 1951 and 1957.
2. Latterly a proportion of cartridges containing very slightly *enriched uranium* was introduced to provide extra reactivity in compensation for the increasing load of neutron-absorbing isotope cans being irradiated in the piles.

BURST CARTRIDGES

3. Out of $\frac{3}{4}$ million cartridges, only 577 actual or suspected bursts occurred, the maximum number being in 1956. The annual totals were:

Year	Bursts
1951	3
1952	10
1953	63
1954	77
1955	126
1956	160
1957	138 (to October)

4. A burst cartridge was not a drastic failure; it was by definition a cartridge which gave an abnormally high signal on the very sensitive BCDG, although detailed examination sometimes failed to reveal any obvious rupture or perceptible hole.
5. Factors considered in cartridge failures were:

    (a) the design and quality of the can;
    (b) the lining;
    (c) the standard of manufacture;
    (d) the behaviour of the can at elevated temperatures;

(e) the behaviour of the can during thermal cycling;
(f) the effect on the can of the period of irradiation.

6. To try to find some way of reducing the number of bursts, the two piles were operated at different maximum temperatures for certain periods in 1953 and 1955. Experience of pile operation (see para. 8 below) indicated that the detected burst rate was not affected by rates of heating or cooling, and it also appeared that (below an obvious limit – far in excess of operating temperatures – at which the aluminium can would become dangerously soft) temperature was not the main factor in causing bursts. The extent of irradiation seemed more likely to be significant, and failures definitely occurred as a result of irradiation growth and/or uranium/aluminium incompatibility. The burst rate was independent of operating temperature, below about 250MWD/t irradiation, but above that the effect of temperature was complex and (depending on heat transfer conditions) catastrophic oxidation could be initiated at low temperatures (see para. 10).

7. Safety therefore depended on being able to detect and discharge an oxidising element quickly, before burning spread to other cartridges or to the graphite moderator.

MAXIMUM OPERATING TEMPERATURES

8. By early 1955 the piles were normally operated at a maximum central metal temperature below 385°C. In July 1955 this was raised to 395°C for Pile No.1, but in Pile No.2 was reduced to 350°C to see if the number of failed cartridges was affected. In the next three months, however, there were more burst cartridges in Pile No.2 than in No.1.

9. Maximum operating temperatures varied from 280°C (1951) to 395°C (1956).

10. The Fleck Committee's cartridge working party concluded after the accident that there was no intrinsically safe maximum temperature, but that a maximum fuel temperature of 395°C was safe in normal operation, with blower cooling (but see para. 13). The Fleck TEC report (Cmnd 471) accepted the existing design of uranium fuel cartridge as satisfactory.

MAXIMUM TEMPERATURES IN WIGNER ANNEALS

11. In Pile No.1, the maximum uranium temperatures recorded in Wigner anneals were:

July 1954	318°C
March 1955	360°C
November 1955	295°C
February 1956	395°C
April 1956	370°C
November 1956	370°C

12. In Pile No.2, the temperatures were similar, ranging from 251°C to 390°C. Thus the maximum fuel element temperatures during Wigner releases were similar to the maximum operating temperatures.
13. The cartridge working party thought that the maximum fuel temperature that would be safe in normal operation might be unsafe in a shutdown reactor, without the blowers and with BCDG not functioning.

Appendix VI: Note on Other Cartridges in the Windscale Piles

LITHIUM-MAGNESIUM (OR AM) CARTRIDGES

The tritium project

1. The project to supply tritium to Aldermaston was initiated in 1954. The tritium was to be produced in the two Windscale piles by irradiating cans containing rods of lithium-magnesium alloy. The project was considered by several special meetings, and also by the Thermal Reactor Cartridge Working Party, the Thermal Reactor Design Committee and an AM Cartridge Working Party.
2. The design of AM cartridge was to assume a maximum uranium metal temperature in the pile of 370°C. A temperature of 350°C for one day a year during Wigner releases could, it was thought, be tolerated. The metallurgist Dr A. B. McIntosh warned that the compatibility of aluminium with magnesium was poor at 400°C because of eutectics: operating at that temperature or above would be 'equivalent to starting with-out a can', he said.
3. In all the early discussions 350°C was assumed as the maximum AM temperature during Wigner releases. There was no recorded discussion of a possible fire hazard in the event of can failure and consequent exposure of AM alloy to air at high temperatures. However, I believe that some Windscale scientists disliked the use of the lithium-magnesium alloy because of the magnesium fire hazard, and would have preferred pure lithium if it had been available.
4. The AM working party last met in February 1955, and the Thermal Reactor Cartridge Working Party last considered AM cartridges in the same month. Direction of the project then passed to the Windscale Technical Committee and the Works General Manager's Committee.

AM cartridge designs

5. Three marks of AM cartridge were produced.

    (a) Mark I   The original design was a bare rod of alloy of 0.5-inch diameter in a standard aluminium isotope can, with weighting pieces of lead metal.

    (b) Mark II   Used for the main production programme, this consisted of a bare AM alloy rod of 0.63-inch diameter canned in aluminium, enclosed in a $\frac{1}{4}$-inch thick lead annulus, the

Other Cartridges in Windscale Piles 173

whole in an outer can of aluminium. The lead layer was known to retard the penetration of aluminium by the alloy at temperatures above 430°C. Large quantities of these cartridges were used in both Windscale piles and went through Wigner releases without evidence of reaction or penetration.

(c) Mark III This consisted of 1-inch diameter rods of bare alloy in an aluminium can, with no lead annulus or second (outer) aluminium can. This type was designed, fabricated and loaded at short notice to increase tritium output when it became clear that the production programme could not be met with the Mark II cartridge.

6. Mark III cartridges were first loaded in Pile No.1 on 4 December 1956, and so had not been through a Wigner anneal in Pile No.1 before the October 1957 accident. However, this Mark had been in Pile No.2 during its anneal of July 1957.
7. A Mark IV design was already under consideration in October 1957. It consisted of a 0.8-inch alloy rod in an inner aluminium can, inside a lead annulus and an outer aluminium sheath. Thus it was similar to Mark II, only larger and heavier.

Post-accident investigations

8. After the October 1957 accident, the R & DB (W) carried out a programme of investigation into possible causes of the fire (abnormal release of Wigner energy to form a hot spot, reaction between graphite and air, AM cartridge catching fire, fuel element catching fire, combustion of materials other than AM in isotope channels).
9. Experimental heating of active AM cartridges from Pile No.2 did not, it was found, develop a gas pressure sufficient to rupture the can. However, in a higher temperature Wigner release, or at a higher rate of energy release, the stage could possibly have been reached where the bursting pressure of the aluminium can was exceeded.
10. Tests on Mark III AM cartridges gave failure rates as follows:

    550°C   All failed in 30 minutes. One ignited.
    450°C   All failed in 1–10 hours. Some caught fire.
    440°C   All failed after 34 hours.
    425°C   Intact after 59 hours.

11. Given a temperature above 440°C, the tests showed that it was only a matter of time before penetration of the can, and a magnesium fire, occurred. During the October 1957 Wigner release one thermocouple recorded above 400°C for 17 hours, and above 420°C for $3\frac{1}{2}$ hours. However, since much of the pile was not served by thermocouples, there might well have been an unrecorded hot spot of over 440–450°C which would have fulfilled the conditions for an AM can failure and fire.
12. When in July 1957 the Mark III cans went through an uneventful Wigner

release in Pile No.2, the maximum recorded graphite temperature had not exceeded 305°C. Later, however, when AM cartridges were removed from the closed-down Pile No.2, some were found to be severely damaged, presumably during that anneal.

13. R & DB (W) made a report on AM cartridges in November 1957, after the date of the Penney report and the White Paper (Cmnd 302). (Some at least of their work had apparently been available to the Penney Inquiry while it was taking evidence.)

14. The R & DB (W) report's conclusions are listed below.

    (a) AM cartridges did not ignite under 400°C in the conditions investigated and even when ignited the reaction tended to be stifled by its products. (The report, RSR 572, considered that the differences of design of AM cartridge were of secondary importance, compared to time/temperature conditions.)
    (b) The cartridges showed considerable individual variations in reactivity.
    (c) The conditions for maximum effect of ignition consisted of a period in relatively stagnant air at about 436°C, followed by a blast of air preferably at a high temperature.
    (d) Under the conditions of the experiment, ignition of the cartridges had little effect on the temperature or the rate of oxidation of the surrounding graphite.

15. The Cartridge Working Party of the Fleck TEC noted that AM cartridges of the Mark III type created a serious problem of compatibility of the metals used, so that at moderately high temperatures the cartridges could deteriorate severely and might oxidise rapidly, or even catch fire, at temperatures as low as 350°C or 400°C. The Fleck TEC report (Cmnd 471) did not discuss the AM cartridges but concluded that it would be 'necessary to modify the existing design of lithium-magnesium cartridge' which it 'considered to be unsatisfactory'.

ISOTOPE CARTRIDGES

16. Other cartridges – apart from the fuel elements and AM cartridges – in the Windscale piles were:

    (a) bismuth oxide cartridges, being irradiated to produce polonium-210, at that time required by Aldermaston for initiators (alpha-emitters used to start the fast chain reaction in plutonium bombs);
    (b) various isotope cartridges being irradiated to produce radioisotopes (especially carbon-14) required for civil purposes by Harwell and the Radiochemical Centre at Amersham.

17. None of these cartridges was thought to have played any part in the Windscale accident.

Appendix VII: Emergency Site Procedure at Windscale

The following outline of the Windscale emergency procedure was sent by Sir Leonard Owen to Sir Edwin Plowden on 16 October 1957.

1. *Classification*
   The emergency conditions which may arise on the site will be classified into two categories:-
   1.1 *Severe*
   One which requires the whole site to be alerted and which may lead to evacuation of whole or part of the Works.
   1.2 *Most Severe*
   One which may lead to evacuation of part of the surrounding countryside as well as the whole or part of the Works. This category may arise from a severe accident and will be declared if the assessed extent of the severe incident justifies it.
2. *Definition*
   The circumstances which may lead to either a *Most Severe* or *Severe* Site Emergency are defined as follows:-

Location	Nature of Hazard	Incident
I Pile Group	A dangerous level of airborne radioactivity	A confirmed 200 curie signal on the stack filter activity meter

(There are detailed instructions on the same lines for all the other locations at Windscale.)

3. *Control during a Site Emergency*
   The control during a Site Emergency, will be exercised by the Works General Manager or, in his absence, by his deputy.
   During the silent hours, the Shift Manager will be authorised to initiate Site Emergency action. The Works General Manager, Works Manager, Works Engineer, and Manager Health Physics and Safety will be informed immediately of the incident.
   Decision on the classification of the incident, *Severe* or *Most Severe*, will be made by the Works General Manager or, in his absence, by the deputy. If the Site Emergency is classified as *Most Severe* the necessary contact with the Local Civil Authorities will be made by the Works General Manager or his deputy.

4. Action by all Personnel

(This section of the emergency procedure consists of detailed instructions to site personnel as to action to be taken to handle the incident; it does not deal with external contacts and is not therefore reproduced.)

5. End of Site Emergency

The end of the Site Emergency will be indicated by the sounding of a series of 10-second blasts on the siren.

Note on Para 5 When an incident occurs involving the spread of radioactivity within and/or without the factory perimeter, the procedure is to inform the local Medical officer of Health, the local police, and the Factory Inspectorate. The Chief Constable has an agreed detailed drill as to the steps he should take to warn the local population.

(The arrangements had been agreed with the authorities concerned at a meeting in Carlisle on 22 November 1954 (see below). Windscale did a detailed survey and listed the names and addresses of the people who might be affected by a district emergency.

MAJOR EMERGENCY AT SELLAFIELD

Note made following a Meeting at The Courts, Carlisle
at 2 pm on 22 November 1954

Present: Mr Swift
The Chief Constable
County Medical Officer
County Surveyor
County Welfare Officer
County Fire Officer
Director of Education
Mr Davey

Summary

In the event of a major emergency at Sellafield, which is a remote possibility, it is not anticipated there will be an explosion with immediate disrupting effects – the type of explosion usually thought of by laymen.

If there is emission of radioactive particles, which are not controlled by the filters, these will reach the air via the chimneys and will travel in the nature of a plume or corridor, their direction being determined by the prevailing wind. The heavier particles will fall nearest the site, but particles of importance to mankind could travel 2 to 3 miles, forming a kind of corridor. The prevailing wind is South West, therefore of concern to Calderbridge, or variation of direction might affect Beckermet, Gosforth or Seascale.

To persons living in the plume, concentrations might be sufficiently low to be met by them staying indoors, with doors and windows closed.

If not, evacuation would be necessary.

It is thought that a warning of two to eight hours may be available.

ACTION TO BE TAKEN AT SELLAFIELD WHEN STAGE HAS BEEN REACHED REQUIRING ASSISTANCE
General Manager will notify Chief Constable or Superintendent of Police at Whitehaven.

Sellafield will supply garments and masks to be used by personnel warning inhabitants.

Direction and area of plume will be stated, indicating area to be warned:

(a) to remain indoors with doors and windows closed.
(b) evacuation.

Staff from Sellafield will be detailed to assist in warning. Lists of addresses prepared at Sellafield will be on hand to pass to personnel concerned, set out in specified areas.

If hard rations are required for schoolchildren at village concerned, these will be issued from Sellafield.

ACTION AT WHITEHAVEN DIVISIONAL HEADQUARTERS
Notify Chief Constable so that the County Medical Officer, County Welfare Officer, County Surveyor may be alerted.

Turn out all available personnel and transport to assist in warning civilians of (a) or (b) as required.

Appendix VIII: Calculations of Emergency Levels for Iodine-131

GENERAL

In 1956, before the Windscale fire, the question of an action level for radio-iodine in milk was twice considered, but no decision was reached on a level for nuclear accidents. (Radioiodine refers to all the radioactive isotopes of iodine. In a reactor accident the iodine released is predominantly iodine-131 and other isotopes can be ignored. In an atomic bomb explosion, other isotopes are plentiful and have to be taken into account. Only one of the instances below (see (1)) is concerned with nuclear explosions. Immediately after the fire an urgent decision had to be made (on 12 October) about local milk supplies. That decision was re-examined on 15 and 19 October and reported in November 1957 in the White Paper on the accident (Cmnd 302). In 1959 the matter was reconsidered by an MRC Committee and its recommendations were published in April 1959 (*British Medical Journal*, vol. 1, pp. 967–9) and later in a White Paper (Cmnd 1225, December 1960).

In each case the radiation dose to young children's thyroids was recognised as the limiting factor because of the child's relatively large milk consumption and small thyroid. (Equal amounts of iodine-131 in a child's small thyroid and an adult's large one will be more concentrated in the former, and thus the dose will be higher.) The known carcinogenic dose to the thyroid was 200–250 rads.

An action level must therefore be based on a maximum acceptable iodine-131 dose to the child's thyroid. But for protective purposes derived limits are needed, whether for concentration in milk or deposition on grazing land. Calculating derived limits involves many factors: the half-life of iodine-131; how much of the deposited iodine-131 is intercepted by the grass, how long it remains there, and how much grass is eaten by the cow; how much of what she ingests is concentrated in her milk; how much milk the child drinks; how much iodine-131 from the milk is retained in the child's thyroid; the size of the thyroid; and the total radiation dose the iodine-131 delivers before it decays.

All these factors are relevant in calculating the integrated radiation dose to the child's thyroid and the appropriate derived action levels.

In each of the approaches summarised below as (2) to (8), it was axiomatic that in peacetime the integrated dose to the thyroids of young children must not exceed 25 or 30 rads.

The first two approaches expressed the limit in terms of contamination of pasture by iodine-131; the other six in terms of radioiodine in milk. The limits are all quite similar, but the underlying calculations are difficult to compare because the successive steps, the assumptions about milk intake, thyroid weight or iodine-131 retention and so on are not always stated.

RADIOIODINE IN MILK: EIGHT APPROACHES TO EMERGENCY ACTION LEVELS

(1) 1956: ARC

An ARC paper of 1956 (A.E.R.E. ARC/RBC 5: R. Scott Russell, R. P. Martin and G. Wortley 'An assessment of hazards resulting from the ingestion of fallout by grazing animals' (1956)) considered conditions in a nuclear war and was not concerned with nuclear accidents. Assessing the hazards of fallout from a nuclear attack, it concluded that a major risk would be to infants drinking milk contaminated by radioiodine. In these conditions it proposed an integrated dose of 200 rads to the child's thyroid as the maximum allowable. Such a dose, it calculated, would result from a concentration of 0.39 microcurie per litre of milk, which would arise from an initial ground contamination level of 3.9 microcurie per square metre, and a daily intake by the cow of 241 microcuries.

This ARC limit, recommended for civil defence purposes, was later confirmed by the MRC as an acceptable risk for a few weeks in wartime conditions, if it was a question of subsisting on contaminated milk or going without. It was *not* considered appropriate in peacetime when alternative milk supplies were available.

(2) 1956: MAFF

The possibility, however remote, of a major nuclear accident prompted MAFF in late 1956 to suggest an emergency limit for radioactive contamination of agricultural land.

MAFF wanted to ensure that in the event of an accident prompt decisions could be made about the control of foodstuffs, including milk, on the basis of ground monitoring. MAFF therefore needed a limit expressed in terms of radioactivity per square metre of land.

Their starting point was the maximum permissible body burden of iodine-131 recently recommended by the ICRP for the thyroid of the adult radiation worker. This was 0.6 microcurie. The MAFF memorandum suggested 0.045 microcurie for the child, presumably assuming a child's thyroid to be 3/40 of the weight of the adult thyroid.

The MAFF scientists calculated that 0.045 microcurie of iodine-131 in the child's thyroid would result, through the food chain, from ground deposition of 7 microcuries per square metre.

Comparison with other calculations is difficult, since the underlying assumptions (e.g., about the pasture–cow–milk pathway, the daily milk intake, or the proportion of ingested iodine-131 retained in the child's thyroid) were not explicitly stated.

It may be noted that later experimental work (R. J. Garner, 'The assessment of the quantity of fission products likely to be found in milk in the event of aerial contamination of agricultural land, *Nature*, 186 (1960), p. 1063) showed that only 1 microcurie of iodine-131 deposited per square metre of ground could result in 0.16 microcurie per litre of milk within a few days. MAFF's paper may have under-estimated the transfer from ground deposition to milk and hence its proposed limit may have been too high by a factor of at least 7.

However, the matter was left unfinished and no agreement was reached on MAFF's suggested action level.

(3) 12 October 1957: AEA

Immediately after the Windscale accident, Authority staff (McLean *et al.*) had to decide quickly what limits on milk were needed to protect the public. They argued as follows (see Dunster, Howells and Templeton, paper on district surveys given at the UN conference on peaceful uses of atomic energy, 1958):

(a) there was medical evidence of thyroid cancers in children after radiation doses of 200 rads, but not at lower levels;

(b) milk supplies to the whole population should be controlled so as to avoid thyroid doses in excess of 20 rads to young children (A. S. McLean had quoted 10 rads, not 20, in explaining the 0.1 microcurie per litre decision at the meeting with consultants on 15 October 1957: see p. 62);
this necessitated calculating the concentration of iodine-131 in milk that would deliver 20 rads to a child's thyroid;

(c) ICRP constants (1953) indicated that 1 microcurie of iodine-131 would deliver a dose of 130 rads per gram of thyroid. In an adult thyroid weighing, say, 20g the total dose would be 130/20 or 6.5 rads; in a child's thyroid weighing, say, 5g the total dose would be more concentrated, namely 130/5 or 26 rads;

(d) a child was assumed to drink 1 litre of fresh milk daily, and to retain 45 per cent of the iodine-131 content in the thyroid;

(e) allowing for the radioactive decay of iodine-131 (half-life of 8 days) the concentration in milk which would expose the child's thyroid to a total of 10 rads was 0.15 microcurie per litre;

(f) alternatively, if it was assumed that the child's thyroid weighed 1.5g, and 20 per cent of the iodine-131 was retained, a 10 rad dose would result from a concentration of 0.1 microcurie iodine-131 per litre of milk.

This figure of 0.1 microcurie per litre was the limit actually adopted for the whole duration of the milk ban.

(4) 15 October 1957: the AEA's radiological consultants

A few days later the AEA's radiological consultants reviewed the calculations of 12 October (see (3) above), together with alternative calculations by Dr (later Sir Edward) Pochin.

The latter assumed:

(a) that the child consumed $\frac{1}{2}$ litre (not 1 litre) of milk daily;
(b) that the child's thyroid weighed 5g;
(c) that 45 per cent of the iodine-131 ingested in milk was retained in the thyroid.

On this basis an iodine-131 concentration of 0.1 microcurie per litre of milk would give an average dose of 7 *(not 10) rads* to the child thyroid, but with a possibility of 20 rads in spots.

The consultants agreed that the AEA action level of 0.1 microcurie per litre was appropriate to the emergency.

(5) 19 October 1957: MRC

Five days later an MRC committee considered the previous calculations (see (3) and (4)). They referred to a carcinogenic dose of 250, not 200, rads. They thought it prudent to assume a daily milk intake of 2 pints (roughly equal to the McLean figure and nearly double the Pochin figure). Their own calculations are not recorded but their conclusion was that in future the action level should be not more than 0.06 microcurie iodine-131 per litre of milk.

(6) 6 November 1957: MRC

An MRC report was published in November as part of the White Paper on the Windscale accident (Cmnd 302, Annex III). It confirmed the original limit (see (3) above) of 0.1 microcurie iodine-131 per litre of milk as 'sufficiently correct for the particular situation'. It did not refer to a 0.06 microcurie limit for the future.

(7) 11 April 1959: MRC

A later report by an MRC Committee, on the maximum permissible dietary contamination after a nuclear reactor accident, was published in the *British Medical Journal* on 11 April 1959 (vol. 1, pp. 967–9), and subsequently reproduced in a 1960 White Paper (*Second Report to the Medical Research Council on Hazards to Man of Nuclear and Allied Radiations*, Cmnd 1225, December 1960, Appendix J).

It recommended that the mean radiation dose to the infant thyroid should not be permitted to exceed *25 rads*. To ensure this, it calculated the maximum permissible concentration of iodine-131 in milk as follows:

(a) the maximum intake of iodine-131 depends on: thyroid size; and iodine uptake from food at different ages;
(b) the weight of the thyroid is taken to be:

child from birth to 6 months (little increase in weight of gland)	1.8g
child of 3 years	3.4g
child of 10 years	9.2g
adult	25.0g

(c) the uptake of iodine-131 is expected to fall from 50 per cent in childhood to 30 per cent in adult life (for small children fed on liquid milk only, the maximum thyroid irradiation would occur at about 6 months);
(d) assuming:

1.8g thyroid weight
50 per cent iodine-131 uptake
intake of 0.9 litre of milk a day
concentration of 0.065 microcurie iodine-131 per litre of milk

then a 6-month child would receive an integrated dose to the thyroid of 25 rads.

(8) 1975: MRC

A 1975 MRC publication ('Criteria for controlling radiation doses to the public after an accidental escape of radioactive material') made the following assumptions:

maximum permissible dose to child thyroid	30 rem
weight of thyroid	1.8g
uptake to thyroid	34 per cent
effective half-life of iodine-131 in thyroid	6 days
dose to thyroid per microcurie ingested	17 rads

The effect of the MRC calculations was to increase the maximum permissible concentrations of iodine-131 in milk to 0.256 microcurie per litre, four times the 1959 limit (see (7) above), and $2\frac{1}{2}$ times the limit applied after the 1957 Windscale fire.

The 1975 maximum permissible concentration is about 5/8 of the ARC's 1956 civil defence figure (see (1) above). However, the maximum dose to a child's thyroid is more than 6 times lower (30 rem compared with 200).

Table A.2 Action Levels for iodine-131 based on maximum permissible exposure of the child thyroid

	ARC 1956 (1)	MAFF 1956 (2)	AEA 12/10/1957 (3)	Consultants 15/10/1957 (4)	MRC 19/10/1957 (5)	MRC Nov. 1957 (6)	MRC 1959 (7)	MRC 1975 (8)
Assumed daily intake of milk by child (in litres except column (5))	1		1	0.5	2 pints		0.9	
Assumed weight of child's thyroid (in grams)	1.5		1.5	5			1.8	1.8
Assumed proportion of ingested iodine-131 retained in child's thyroid (%)	20		20 (or 45)	45			50	34
Iodine-131 content of child thyroid (in microcuries)		0.045						
Integrated dose to child thyroid (in rem)	200		10 (or 20)	7			25	30
Action level for milk (microcuries per litre)	0.39		0.1	0.1	0.06	0.1	0.06 (f)	0.256
Action level in terms of ground contamination (microcuries per m²)	3.9	7						

Notes
(a) The radiation units are those in use at the time (rems, millicuries and microcuries)
(b) Columns (1) to (8) correspond to sections (1) to (8) in the text of Appendix IX
(c) Column (1) relates to nuclear war conditions, columns (2) to (8) to peacetime nuclear accidents
(d) Iodine-131 is predominant in peacetime accidents; other isotopes of iodine are taken into account in Column (1)
(e) The carcinogenic dose to the thyroid was generally assumed to be 200–250 rads
(f) If children under 3 were fed on dried milk or milk from an uncontaminated source the limit could be raised to 0.3 microcurie/litre

Appendix IX: Estimates of Fission Product and Other Radioactive Releases Resulting from the 1957 Fire in Windscale Pile No.1

1958

Iodine-131	20 000Ci
Caesium-137	600Ci
Strontium-89	80Ci
Strontium-90	9Ci

Smaller quantities of other fission products such as ruthenium-103 and ruthenium-106, zirconium-95, niobium-95 and cerium-144, together with polonium-210, were also released.
(*Source*: PUAE PB16/UK 'District Surveys following the Windscale accident 1957' H. J. Dunster, H. Howells and W. L. Templeton at the 1958 Geneva Conference on Peaceful Uses of Atomic Energy).

1960

Iodine-131	20 000Ci
Tellurium-132	12 000Ci
Caesium-137	600Ci
Strontium-89	80Ci
Strontium-90	2Ci

(*Source*: Appendix H in *Second Report to the Medical Research Council on Hazards to Man of Nuclear and Allied Radiations*, Cmnd 1225, HMSO, December 1960).

Table A.3 Estimates of total activity in damaged fuel, activity retained in filters, and activity released to atmosphere

Radioactive Substance	Activity in damaged fuel (ci)	Activity retained in filters (ci)	Activity released to atmosphere (ci)	Released to atmosphere (%)	Retained (%)
Iodine–131	168 000	30 000	20 000	12.0	73
Tellurium–132	160 000	20 000	12 000	7.5	92

Table A.3 continued

Radioactive Substance	Activity in damaged fuel (ci)	Activity retained in filters (ci)	Activity released to atmosphere (ci)	Released to atmosphere (%)	Retained (%)
Caesium–137	8 000	80–1 000	600	7.5	80
Strontium–89	380 000	80–560	80	0.02	99
Strontium–90	9.200	0.12–1.2	2	0.02	99
Ruthenium–106	12 000	40–140	80	0.7	98
Cerium–144	290 000	8–84	80	0.03	98

(*Source*: Data taken from J. R. Beattie, 'An assessment of environmental hazard from fission product releases' (AHSB(S) R64, 1963)).

1974–1982

Various reassessments of the total release were made between 1974 and 1982: Clarke (see source to Table A.4) gave a lower figure for iodine-131 (16 200Ci) but somewhat higher figures for other fission products, and listed a larger spectrum of nuclides (not shown here) than previously. A. C. Chamberlain (see source to Table A.4) calculated the iodine-131 release as 27 000Ci, and estimated that a few hundred curies of polonium-210 and perhaps 100 000 Ci of tritium were released. Figures were given by P. J. Taylor, in 'The Windscale Fire, October 1957', report for the Union of Concerned Scientists, Cambridge Mass. (PERG 1981), of other radionuclides, in particular krypton-85, yttrium-90, zirconium-95, niobium-95, ruthenium-103, tellurium-129, xenon-131, xenon-133 and xenon-135, barium-140, molybdenium-99 and cerium-141.

Table A.4 Summary of estimates of release 1958–82 (in ci)

	Dunster Howells and Templeton, 1958	Cmnd 1225, 1960	Beattie, 1963	Clarke, 1974	Chamberlain, 1981
Iodine–131	20 000	20 000	20 000	16 200	27 000
Caesium–137	600	600	600	1 230	?
Strontium–89	80	80	80	80	200
Strontium–90	9	2	2	6	?
Tellurium–132	–	12 000	12 000	16 100	?
Ruthenium–106	–	–	80	160	?
Cerium–144	–	–	80	109	?

continued on page 186

Table A.4 continued

	Dunster Howells and Templeton, 1958	Cmnd 1225, 1960	Beattie, 1963	Clarke, 1974	Chamberlain, 1981
Polonium–210	–	–	–	–	A few hundred
Tritium	–	–	–	–	?100 000

Source: J. R. Beattie, 'An assessment of environmental hazard from fission product releases', UKAEA, AHSB(S) R64 (1963); Cmnd 1225, *Second Report to the Medical Research Council on the Hazards to Man of Nuclear and Allied Radiations* (HMSO 1960); A. C. Chamberlain, 'Emission of fission products and other activities during the accident to Windscale Pile No.1 in October 1957', AERE-M3194 (July 1981); R. H. Clarke, 'An analysis of the 1957 Windscale accident using the WEERIE Code', *Annals of Nuclear Science and Engineering*, 1 (1974), pp. 73–82; H. J. Dunster, H. Howells and W. L. Templeton, 'District surveys following the Windscale accident, October 1957', *Proceedings of the 2ⁿᵈ UN Conference on the Peaceful Uses of Atomic Energy*, 18 (1958), pp. 296–308.

Appendix X: Estimates of Total Radiological Impact of the Radioactive Releases Resulting from the 1957 Fire in Windscale Pile No.1

Table A.5 Estimates of total radiological impact of 1957 fire

	Estimated maximum number of non-fatal cancer cases (UK)	Estimated maximum number of cancer deaths (UK)
November 1957 (Cmnd 302)	0	0
December 1960 (Cmnd 1225)	0	0
July 1981 (Taylor)	248	10–20
1982 (Crick and Linsley)	237	20
September 1983 (Crick and Linsley)	?	35
1988 (Clarke)	90 (plus 10 hereditary defects)*	100*

* But note that these are upper bound figures and Clarke commented on them: 'At the very low levels of individual doses involved it must be questionable as to whether the risks are real ... The most likely number of health effects will be lower and may be zero.'

Sources: Cmnd 302, *Accident at Windscale No.1 Pile on 10 October 1957* (HMSO, 1957); Cmnd 1225, *Second Report to the Medical Research Council on the Hazards to Man of Nuclear and Allied Radiations* (HMSO, 1960); P. J. Taylor, 'The Windscale fire, October 1957', report for the Union of Concerned Scientists, Cambridge, Mass. (PERG 1981); M. J. Crick and G. S. Linsley, 'An assessment of the radiological impact of the Windscale reactor fire, October 1957', NRPB R135 (1982), and NRPB R135 Addendum (1983); R. H. Clarke, 'The 1957 Windscale accident revisited', International Conference on the Medical Basis for Radiation Accident Preparedness - II Clinical Experience and follow-up since 1979: Oak Ridge, Tennessee, 20–22 October, 1988.

Appendix XI: Report of Penney Inquiry

REPORT ON THE ACCIDENT AT WINDSCALE NO.1 PILE
on 10th OCTOBER, 1957

To the Chairman, U.K.A.E.A.

CHAPTER I. INTRODUCTION

1. You appointed the Committee of Enquiry on the 15th October with the following Terms of Reference: 'To investigate the cause of the accident at Windscale No.1. Pile on 10th October, 1957, and the measures taken to deal with it and its consequences; and to report'.

2. The Committee sat at Windscale Works from 17th to 26th October inclusive, with a break of one day, during which time evidence was received from the Witnesses whose names are listed at Appendix A. [not reproduced here] The majority of these Witnesses were called by the Committee, but in addition an invitation was extended to any Windscale Works personnel to appear before the Committee if they had information to offer pertinent to the accident. Four such Witnesses came forward (Mr. Evans, Dr. Leslie, Mr. Shackcloth and Mr. Mowat). In addition we received written evidence in the form of graphs, reports, minutes of meetings, etc.* Some of these represented the results of calculations carried out during the Enquiry at the Committee's request.

3. We visited No.1. Pile, the scene of the accident, and also Pile No.II which was shut down in order that we could examine a Pile in shut-down conditions such as prevailed in No.1. Pile before the accident. In addition, we visited the Health Physics Centre to inspect the records of monitoring surveys which had been carried out outside the Factory.

4. Our report is presented in the following sections:

Chapter I	Introduction
Chapter II	Events Leading up to the Accident
Chapter III	Cause of the Accident
Chapter IV	The Measures Taken to deal with the Accident
Chapter V	The Measures Taken to Protect the Workers
Chapter VI	The Measures Taken to Protect the Public
Chapter VII	Conclusions
Chapter VIII	Recommendations.

* As this report was required urgently, it has not been possible within the time available to digest this material into properly edited Appendices. Arrangements have, however, been made for its preservation (and see para.111 (a)).

CHAPTER II. EVENTS LEADING UP TO THE ACCIDENT

5. The accident occurred in the course of a controlled release of stored Wigner energy from the graphite.

6. The first time that Wigner energy was released in the Windscale piles occurred spontaneously in September, 1952. As a result of a study of this accident, a procedure was instituted for controlling releases of Wigner energy. Over the period to the end of 1956, eight such releases had been carried out on Pile No.I [sic]. The general procedure has been to shut down the Pile, to arrange the appropriate instrumentation and then to cause the Pile to diverge with no coolant air flow, thus raising the graphite and uranium temperatures. By this means, the graphite is brought up to a temperature at which Wigner release is started. This has not always succeeded in annealing all the graphite in the Pile, e.g. in 1956 one attempt (in April) was completely unsuccessful and two others partially successful; that is to say, energy releases were recorded from certain regions only of the Pile and pockets were left un-annealed.

7. Originally the procedure was to carry out Wigner releases after 20,000 cumulative megawatt days. Later this figure was increased to 30,000 megawatt days. The Windscale Works Technical Committee in September, 1957, considered a paper (Ref. IGR-TN/W.586) which recommended an increase to 50,000 cumulative megawatt days in view of the increasing difficulties of obtaining a successful release. Until more experience had been obtained, it was decided that the next release should be at 40,000 megawatt days. A Wigner release therefore became due in October, 1957.

8. A note on the loading of the Pile is relevant at this stage. The Pile consists of a structure of graphite blocks pierced by horizontal channels on an $8\frac{1}{4}"$ square lattice pitch. Each channel contains a string of uranium fuel elements. For purposes of charge and discharge the channels are arranged in groups of four, access to each group being by way of a charge hole in the front shield. In the middle of each of the groups is a channel of smaller diameter normally used for the irradiation of isotope cartridges.

9. At the material time, the Pile contained natural uranium and slightly enriched (1.28 Co) uranium in the fuel cartridges, and a considerable variety of isotope cartridges.

10. There are vertical channels in the Pile used for experiments connected with the civil reactor programme. At the time of the accident, all these channels were empty except one which contained a small magnet under test.

11. On 7th October at 01.13 the Pile was shut down and the main blowers switched off in preparation for the Wigner release. All necessary steps were taken to verify that the Pile was completely shut down. The thermocouples used for following Wigner releases were then checked, and those which were unserviceable were replaced.[*]

12. The shut down fans were switched off and the door in the base of the chimney and the back inspection holes on the Pile roof were opened to

---

[*] The location of the uranium and graphite thermocouples is shown at Appendix B [not reproduced here].

minimise coolant air flow through the Pile. At 19.25 the Pile was made to diverge to generate nuclear heat for triggering off Wigner energy.

13. The procedure for Wigner release is to concentrate as much flux and therefore as much heat as possible in the front lower region of the Pile by suitable manipulation of the lower coarse control rods, the upper control rods having been disconnected when fully in.

14. At 19.25 the Pile began to generate nuclear heat and the power level was gradually increased so that, by about 01.00 on 8th October, 1.8 MW was registered on the Pile power meter. (We have established that when the upper horizontal control rods are completely inserted, the ionisation chamber measuring pile power is masked. The power meter reads low. As, however, the Pile was being controlled on temperature recordings, the only effect was to alter the relation between the rod movements and the power changes and hence the 'feel' of the controls).

15. It should be noted at this point that the operation of a Wigner release is the responsibility of the Pile Physicist and his two deputies, by virtue of their specialised knowledge. The Pile Control Engineers operate on the instructions of the Pile Physicists during Wigner releases. It appears that there is nothing in the nature of a Pile Operating Manual. The only written instruction which was produced for our inspection was a minute of 14th November, 1955, a copy of which is at Appendix C [reproduced on p. 167]. However, the Pile Manager informed us that there was a limit of 250°C maximum cartridge temperature which, it had been laid down, should not be exceeded in the first instance during a Wigner release. This temperature was observed in two channels around midnight on 7th October; accordingly, control rods were run in again and the Pile was shut down by 4 a.m.

16. Most of the graphite temperatures rose in the manner normal in Wigner releases. However, according to the Physicist in charge and the Pile Manager, at about 9 a.m. on 8th October the general tendency was for the graphite temperatures to be dropping rather than rising and it seemed probable that unless more nuclear heat was applied, the release would stop.

17. We have carefully examined the thermocouple records on this point. They do not wholly bear out the observation reported in the previous paragraph. Undoubtedly, some of the graphite temperatures were falling and there were some in which no Wigner release was apparent; but a substantial number of the graphite thermocouple readings showed steady increases. We would not therefore endorse the observation that the general tendency was for the graphite temperatures to be dropping rather than rising.

18. However, acting on the observation which he had made, the Physicist in charge decided to boost the release with more nuclear heating. (It should be noted that second nuclear heating had been utilised during three previous Wigner releases. On the first two occasions in 1954 and 1955 it was not utilised until at least 24 hours after the last regional burst of temperature, and after all the graphite temperatures had been seen to be dropping. In 1956 all graphite thermocouples except one were showing a fall).

19. The bottom rods were withdrawn and the Pile made divergent at 11.05 with the object of raising the maximum uranium temperature, which at that time was 300°C, to 350°C. The uranium thermocouple readings show a

sharp temperature increase when the Pile diverged for the second time. The highest uranium temperature recorded was in channel 25/27; the thermocouple reading for this channel rose by 80° to 380°C in a matter of 15 minutes, with a maximum rate of rise of about 30°C a minute. The thermocouple reading was reduced to 334°C within 10 minutes by adjustment of the control rods. Nuclear heating was maintained at a lower level until 1700, and during this period the highest uranium thermocouple readings rose to about 345°C. At 1700 the nuclear heating was terminated.

20. (It should be noted that the positions of the uranium thermocouples correspond to the readings of maximum temperature during normal operation, but that these are not the positions of maximum uranium temperature during a Wigner release. (See also para. 42 below)).

21. During Wednesday, 9th October, the uranium temperatures as recorded show a maximum of 360°C, while at 2200 the highest value recorded was 340°C. The graphite temperatures showed considerable variation, but the general tendency was for the temperatures to increase following the second nuclear heating. One graphite temperature in particular, in channel 20/53, which had shown a reading of about 255°C at the time when the second nuclear heating was applied, continued to rise steadily until by 22.00 it had reached a temperature of 405°.

22. The high temperature being recorded in channel 20/53 caused the Pile Physicist, at 21.00, to shut the chimney base and the inspection holes to allow the chimney draught to induce some flow of air from the Pile and thus cool it. The effects were not considered big enough and at 22.15 the fan dampers were opened to give a positive air flow through the reactor. The dampers were open for 15 minutes on this occasion. They were opened again for 10 minutes at 00.01 on 10th October, for 13 minutes at 02.15 and for 30 minutes at 05.10. This had a cooling effect on all graphite temperatures except 20/53, where the temperature rise was merely arrested.

23. The records from the pile stack activity meter show no special features during the early stages of the operation. There was an almost imperceptible increase in activity before and during the period of the first three damper openings. At 05.40, at the end of the fourth damper opening, there was a sharp increase of six curies. This was noted by the Physicist who was then on duty, but no special action was taken because he regarded it as the normal consequence of the first movement of air through the pile and up the stack. This increase was followed by a steady drop in the curve for about two and a half hours after which time stack activity rose steadily to a figure of 30 curies at 16.30 on 10th October.

24. The outlet duct air temperature had remained stationary at about 40°C until 07.00 on 9th October, but rose steadily thereafter. With peaks occasioned by the successive damper openings it had reached 85°C by 08.00 on 10th October.

25. The graphite temperature in channel 20/53 continued to rise after temporary reductions during the periods of damper opening, until at 12.00 on 10th October a temperature of 428°C was recorded. The dampers were again opened for 15 minutes at 12.10 and for 5 minutes at 13.40. During these openings the second and very large increase in stack activity, which has already been mentioned, was noted. At about the same time, a high activity

reading on the Meterorological [*sic*] Station roof was reported.

26. These effects suggested to the operating staff the existence of one or more burst cartridges. At 13.45 the shut down fans were switched on as a preliminary move in an attempt either to use the scanning gear to detect any burst cartridges or alternatively to blow the Pile cool. At 14.30 the turbo-exhauster was switched on in order to scan for the burst slug. It was then found that the scanning gear was jammed and could not be moved. The Pile Manager informed us that at the end of previous Wigner releases the scanning gear could not be moved, presumably because of over-heating. On the present occasion it was immovable despite the fact that the Maintenance Section had worked on it on the previous day (9th October) and had moved it.

27. Being unable to operate the scanning gear, recourse was had to the MX 119 to sample the air coming from the Pile for particulate activity. This showed a positive large reading.

28. At this stage the Works General Manager was informed by the Pile Manager that there appeared to be a bad burst, and he instructed that the affected channel should be identified and discharged as soon as possible.

29. Since the scanning gear could not be used it was decided to remove the charge plug and inspect the uranium channel showing the highest readings.

30. Certain steps had to be taken before entry could be effected to the charge hoist. An air count had to be taken to ensure that there would be safe working conditions and the operatives themselves had to have a full change into protective clothing. When access had been secured to the charge hoist, a further delay of a few minutes ensued because of incorrect labelling of a thermocouple. The temperature of fuel channel 21/53 had been recording very rapid increases until at 16.30 it was in the neighbourhood of 450°C. At this time, the plug covering this group of four channels was pulled out and the metal was seen to be glowing; the graphite appeared to have its normal colour. No sign of abnormality was observed in the isotope channels, but no positive conclusion can be drawn from this because the annular gap is much less in the isotope channels than in the uranium channels.

31. This is the appropriate point at which to end of the present chapter, as it is clear from our enquiries that the accident had happened by this time; what followed were remedial measures, belonging appropriately to Chapter IV.

CHAPTER III. CAUSE OF THE ACCIDENT

32. We have come to the conclusion that the primary cause of the accident was the second nuclear heating.

33. This was carried out between 11.05 and 17.00 hours on Tuesday, 8th October, when parts of the pile were still rising in temperature. Having regard to the high temperatures already existing in parts of the pile at this time, the second nuclear heating was too soon and too rapidly applied.

34. The rate of heating for a short period from 11.20 to 11.35 was particularly severe and caused a maximum rate of rise of the observed uranium temperatures of approximately 10°C per minute.

35. The second nuclear heating led to the accident through a chain of events, and here we must put the possible consequences in order of prob-

ability. In our opinion, by far the most likely possibility is that the rapid rise of temperature of the fuel elements due to the second nuclear heating caused the failure of one or more of the fuel element cans. The exposed uranium oxidised and gave further release of heat, which, together with the rising temperatures occasioned by later Wigner releases, initiated the fire.

36. A second possibility which we cannot entirely reject is that a Lithium-Magnesium cartridge failed because the second nuclear heating triggered off pockets of Wigner energy at a time when the general level of temperatures throughout the pile was high. The oxidation of the Lithium-Magnesium could have added further heat and initiated the fire.

37. Once a cartridge had failed whether it were Uranium or Lithium-Magnesium, the burning of the graphite would make some addition to the heat being released and would make its contribution to the development of the fire.

38. We have studied and rejected the possibility that the source of the fire was an isotope cartridge other than the Lithium-Magnesium type mentioned above. We have ascertained that the residual heating due to radioactive decay and gamma ray absorption is insignificant, and at most amounts to 3°C.

39. We consider that the evidence points to the initiation of the fire being the result of the failure of the can of a fuel element giving rise to oxidation. We also consider that the origin lay in the region of the Pile just below the middle plane and towards the front.

40. The picture we have formed of this phase of the incident is as follows.

41. The uranium thermocouples were installed in all cases in cartridges about 16 feet from the front face of the Pile where the maximum temperatures occur under normal operating conditions. Since 1954, these thermocouples have been used to control the progress of the nuclear heating for Wigner release.

42. The uranium thermocouple readings were lower than the maximum uranium temperatures elsewhere in the channels for two reasons. Firstly, the control rod positions were arranged to concentrate the nuclear heating in the front lower part of the Pile and we have examined a calculation which indicates that the peak neutron flux under these conditions was some 3 feet nearer the front face of the Pile than under normal operating conditions.

43. Secondly, every effort was made to minimise air flow through the channels to assist the nuclear heating in initiating the Wigner release; the position of maximum uranium temperatures thus correspond to the position of peak neutron flux.

44. The combined effect of the distortion of neutron flux together with the transfer of the point of maximum temperature to coincide with the point of maximum flux resulted in the position of maximum uranium temperature occurring some seven feet nearer the front face of the Pile than the thermocouple positions. We estimate that the maximum temperature rise of the uranium could be 40% greater than the recorded temperature rise during nuclear heating.

45. The second nuclear heating was applied at 11.05 on 8th October when the majority of the graphite temperatures and all the uranium temperatures

in the channels in the affected region were rising and when the maximum recorded uranium temperature in channel 25/57 was 300°C. The rate at which nuclear heating was applied was unusually rapid and led to a recorded rate of uranium temperature rise of about 10°C/minute to a maximum value of 380°C at a point 16 feet from the Pile face. The normal practice in pile operation is to limit such rises to 2°C/minute.

46. We have calculated that at the point of peak neutron flux, where a graphite temperature in the vicinity was measured as 315°C, the uranium temperature could initially have been at 340°C. The maximum uranium temperature after the sudden nuclear heating could thus have been as much as 450°C for some minutes.

47. That this treatment probably led to the immediate bursting of one or more uranium cartridges is deduced from the following.

48. Some of the Mark X fuel cartridges in the affected region had received an average of 287 MWD/tonne. Evidence was given to us that under normal operating conditions in the Windscale piles, bursts of this type of cartridge occur very infrequently at doses below 280 MWD/tonne, but that at higher dose rates bursts become increasingly frequent. With Mark X cartridges, the incidence of bursts has been mainly in the front end of the Pile and many of these bursts have been caused by growth of the uranium rod causing a crack in the shoulder at the cartridge end.

49. We consider that in this way some cartridge bursts could have been produced during the early part of the second nuclear heating by the rapid differential growth of the uranium relative to the aluminium. The ends would have been pushed off cartridges which had already had a dosage which brought them into the state of increasing susceptibility of bursting.

50. Under these conditions, slow oxidation of the exposed uranium metal would occur. Laboratory tests have shown that complete oxidation of a uranium cartridge is possible within a period of 24 hours if the cartridge is held at a temperature of 400°C inside a furnace. With sound undamaged cartridges the number of failures at this temperature would be extremely small, but with burst cartridges oxidation would be inevitable at 400°C and would accelerate rapidly at higher temperatures.

51. It may be noted that the total heat of combustion of the uranium in a complete channel comprising 17 Mark X cartridges and 4 Mark VI cartridges amounts to the very large figure of 50,000 kilogram calories. Once the assumption is made that a single uranium cartridge is ignited, there is no difficulty in visualising several probable routes by which final conflagration is reached. It is to be expected that slow combustion continued during the period of stagnant air conditions until 22.15 on 9th October, when the shut down air dampers were first opened to admit air to the pile.

52. The rate at which air will begin to circulate through the pile on first closing the chimney base and opening air dampers on the inlet side will inevitably be slow. This is due to the fact that the air in the pile will be stagnant, the chimney will be cold, and there will be little pressure difference available to draw air through the core. The first three periods during which the dampers were opened were too short to permit any steady current of air to be established through the pile. For this reason the fission product activity

released from the oxidising fuel elements was not carried to the top of the stack. But on the fourth opening of the dampers (at 05.10 hours on Thursday, 10th October), lasting for 30 minutes, a sufficient flow of air through the pile was established to carry a cloud of gaseous fission products and also some particulate activity up to the filter at the top of the chimney. At 05.40 a sharp increase was recorded on the pile stack activity meter.

53. During this period, the intermittent supply of air to the pile caused an accelerated rate of oxidation of the smouldering uranium fuel elements in the affected zone. Slow combustion of uranium continued during the morning of Thursday, 10th October with locally rising graphite temperature. Switching on the shut-down fans at 13.45 had the effect of rapidly increasing the rate of combustion in the affected channels. By approximately 15.00 hours a serious fire was raging in the neighbourhood of the 20/53 group of channels.

54. The evidence which we obtained on the Lithium-Magnesium cartridges can be summarised as follows:-

55. Of the various marks of can present, the most likely to have given trouble was the Mark III which has a single can holding the alloy in the form of a rod. Laboratory tests have shown that above 427°C penetration of the can is possible as a result of the formation of a eutectic alloy. At 440°C all cartridges tested failed within 34 hours; at 450°C all failed in a few hours and some caught fire.

56. There is evidence from an analysis of particulate matter caught before 15.30 on October 10th by a cyclone filter at the top of the stack that some lithium and magnesium had burnt by then. However, no temperature as high as 400°C was recorded in the graphite structure until 17.00 on Wednesday, 9th, and then only momentarily and in one graphite channel (21/48). For this reason we consider it unlikely that the lithium-magnesium cartridges were the immediate cause of the fire.

57. For a portion of a graphite block to have been ignited at an early stage would also have required temperatures of 450°C. The possibility of a local large release of Wigner energy in a pocket of graphite which had escaped previously annealing has been considered by us in the light of both thermocouple readings, laboratory data and general information. The available data indicate that the oxidation rate of graphite is very slow indeed at temperatures below 400°C, and since we can find no evidence for the likelihood of serious 'pocket' releases or of temperatures in excess of 400°C, we do not accept this as an explanation of the accident.

CHAPTER IV. THE MEASURES TAKEN TO DEAL WITH THE ACCIDENT

58. Immediately glowing metal was seen in the 21/53 group of channels an attempt was made to discharge the fuel cartridges, but they were stuck fast and could not be moved.

59. It was considered inadvisable to switch on the main blowers in an endeavour to reduce temperature; in view of the high stack activity, this would probably have caused a serious neighbourhood hazard. The shut down fans had to be kept on, however, in order to maintain tolerable working conditions on the charge hoist. The workshops were requested,

Report of Penney Inquiry

therefore, to make graphite plugs in order to blank off the over-heated channels. However, it was shortly afterwards found that in addition to the 21/53 group of channels detected by thermocouple reading, there was a rectangular region of some 40 groups of channels – about 150 channels in all – showing red heat. The attempt to make graphite plugs was abandoned, as it was now clear that the very large number required could not have been made in time.

60. From this time – about 17.00 on 10th October – two principal remedies were attempted. Efforts continued throughout the night to discharge channels from the hot region. Early in the morning of 11th October it was necessary to bring scaffolding poles from Calder construction site as the supply of steel push rods was giving out. Some of the hot channels were discharged by this means.

61. Secondly, at about 17.00 on 10th October it was decided to make a fire break by discharging a complete ring of channels around the hot region. This was successfully carried out. Later, a second row of channels was discharged above and at each side of the hot area; and later still, as the fire continued to threaten to spread upwards, a third row was discharged above the hot area. Discharge had to be suspended at one point while skips were moved in order to avoid a criticality hazard in the water duct.

62. Two subsidiary measures were attempted with no success. The use of argon was considered, but it was found that insufficient was available in the works. Secondly, a tanker of Co_2 was brought over from Calder Hall, and arrangements were made for Co_2 to be supplied to the hot channels. Co_2 was fed into channel 20/56 at 04.30 but with no appreciable effect.

63. Meanwhile observation from the top of the pile, through the east inner inspection hole, revealed an obvious glow on the pile rear face at 18.45: at 19.30 the flames were much brighter, at 20.00 they were yellow and at 20.30 they were blue.

64. At about this time the use of water was first considered. Two hazards had to be examined; first the danger of a hydrogen-oxygen explosion which would blow out the filters, second a possible criticality hazard due to replacement of air by water. The Management were informed, however, of the danger of releasing high temperature Wigner energy if the graphite temperatures were to rise much higher than 1,200°C. It was thought that this might well ignite the whole Pile.

65. By about midnight the Works General Manager had decided that if the other measures failed to secure a reduction in temperature, water should be used. This was the major decision and it was supported by the Director of Production and the Deputy Works General Manager. The fire brigade was ordered to stand by with all available pumps, and work started on the preparation of equipment to enable water to be injected into the channels which had been discharged.

66. Shortly after midnight the Chief Constable was warned of the possibility of an emergency and men in the factory were warned of an emergency with instructions to stay indoors and wear face masks.

67. At 01.38 the graphite in channel 20/53, near the top of the 'hot' area, showed a temperature of 1,000°C, and a fuel element temperature of 1,300°C was recorded by optical pyrometer. Over the next two hours, 'brute force'

efforts were successful in discharging nearly all the top row of burning elements, but the fire continued unabated elsewhere.

68. By 03.44 the water hoses were ready to be coupled at 15 minutes notice.

69. Visual inspection at 04.00 through two of the pile roof inspection holes still showed blue flames: the graphite appeared to be burning.

70. After the unsuccessful use of CO_2, temperatures continued to rise. Efforts to discharge the burning cartridges continued, but by 0700 it was clear that the fire was not being checked. At 0700 it was decided that water should be used, but that before it was turned on all factory labour should be under cover. Some delay was therefore necessary while the shift changed over at 08.00. Water was finally turned on at 08.55 and poured through two channels above the maximum height of the fire. From an initial rate of 300 gallons/minute the flow was increased to 800 gallons/minute. No dramatic change resulted; at 09.56 flames were still feathering out of the back of the pile. At 10.10, therefore, the shut down fans were closed off to reduce air flow through the pile. The fire immediately began to subside. At 12.00 two more hoses were installed, and the water flow was increased to 1,000 gallons/minute. The flow continued at this rate until 06.45 on 12th October, and was then gradually reduced until at 15.10 it was completely stopped. By this time the pile was cold.

CHAPTER V. THE MEASURES TAKEN TO PROTECT THE WORKERS

71. On 10th October, between 11.00 and 14.00, a 3-hour air sample was taken outside the Health Physics Administration building. The sample gave a count of 3,000 ß d.p.m./M^3 compared with a normal level of 200 to 300 ß d.p.m./M^3. This was one of the pieces of evidence indicating the probability of a burst cartridge, and this information was passed to the Assistant Works Manager and to the Pile Manager about 14.00. From 14.15, half-hourly air sampling was undertaken over the Factory site, at some 10 to 15 different points.

72. Rising air counts led to an instruction, at 01.33 on 11th October, that workers should stay indoors and should wear face masks. At 02.30, when the activity appeared to be somewhat reduced, workers were instructed to remove face masks and to hold them ready.

73. The air contamination rose to a worrying but not dangerous level during the morning of 11th October. The values were patchy and varied with time. Expressed in terms of a maximum permissible exposure to any fission product in the air, by I.C.R.P. standards, for lifetime breathing, the values recorded rose from about two at 14.00 on 10th October to values in the region of five to ten during the night, with a few exceptional peaks as high as 150 on the following morning. By 12.00 on 11th October the air activity was dropping fast and the value was one to two. The value of 2 was never again exceeded and most of the readings were about $\frac{1}{4}$ or less.

74. The highest readings on the morning of 11 October were around the chemical plants and the Calder site. There was no contamination, however, inside buildings. The chemical plants were shut down and the workers instructed to sit in the main canteen; construction work on Calder B was

stopped and the workers sent home. Earlier, workers had been instructed to stay within buildings at the time when water was turned on.

75. When operations started at the No.1 pile charge hoist, standard procedure for health control was instituted. Special clothing and film badges were provided at a control point, and operatives were not allowed on to the charge hoist without the full protective clothing and personal dosimeters. Special arrangements were made to ensure the rapid monitoring of the operatives involved. In the evening, extra staff were called in to man the surgery to deal with any severe cases of exposure, but there were none.

76. We have examined the total body radiation records of all the workers concerned in the accident. It is necessary to distinguish between exposure during the accident itself and exposure over the standard control period. The I.C.R.P. tolerance level, which was formerly 3.9r per 13 week period, has recently been reduced to 3.0r, and this is the standard now in force at Windscale Works. Over the 13 week period up to 24th October, 1957 (i.e. including the accident) only 14 of the workers concerned in the accident exceeded the maximum permissible level. The highest figure recorded was 4.66r. Records of exposure during the accident itself are not as reliable as the 13 week records, as the latter are taken from film badges, the former from quartz fibre electrometers which give an approximate reading only. According to the Q.F.E. [quartz fibre electrometer] readings, two workers received 4.5r during the accident, one 3.3r, and there were four others in excess of 2r. All the workers who received doses in excess of the maximum permissible level have been taken out of contact with work involving radiation, in accordance with standard procedure.

77. No worker had to be detained after the accident. Some had hair and hand contamination which was successfully removed by the standard decontamination procedure. One worker was not completely decontaminated at the first attempt, and his hands were gloved and his hair covered until the following morning when decontamination was successfully completed.

78. A thyroid iodine survey was made among the workers and is continuing. The I.C.R.P. level for safe continuous and constant iodine activity in the adult gland is 0.1 μ.c. [microcurie]. The highest thyroid iodine activity so far measured among the staff is 0.5 μc. Since iodine has a short life some increase over the I.C.R.P. level can properly be made if the dose occurs on a single occasion.

79. A survey is also being made among the workers for strontium activity, both Sr 89 and Sr 90. The radio-chemical analysis takes time and so far the two isotopes of strontium have not been measured separately. However, by making a reasonable partition of the activity marked between these two isotopes, the first 25 results which have been so far obtained show levels which are at most one-tenth of the maximum permissible body burden.

80. The evidence so far on radioactive caesium is also satisfactory

81. When the plugs were being removed from the charge wall to find the limits of the area of the fire and to eject burning cartridges, some men looked for a few seconds through the open plug holes towards the pile face from a distance of several feet back from the charge wall. A few men were not wearing head film badges, although they were wearing the normal type of

film badge. These men might have had a dose to their heads, somewhere between 0.1 r and 0.5 r, in addition to the whole body dose recorded by their film badges.

82. During the morning of 11th October the Windscale and Calder canteen managers consulted the Medical Department about their food supplies. They were instructed to take a few simple precautions and this they did.

CHAPTER VI. THE MEASURES TAKEN TO PROTECT THE PUBLIC

83. A note of wind directions is necessary to appreciate some of the developments concerning public health. Throughout Thursday, 10th October, the ground wind was light, but mainly off-shore, i.e. N.E. or N.N.E. During the night it changed to N.N.W. and throughout Friday a 10-knot wind blew, mainly N.W. and N.N.W., i.e. down the coast. Still later it appears to have changed to a S.W. direction.

84. There may well have been an inversion at a few hundred feet above ground level during part of the period when radioactivity was escaping through the filter; and the winds above the inversion may not have been in the same direction as the ground wind. The fall-out pattern as it is now known strongly suggests meteorological conditions of this type.

85. At about 15.00 on Thursday, 10th October, a survey van was sent out to make district surveys in the down-wind direction of the ground winds, i.e. along the cinder track towards Seascale. Because it is so much quicker to make gamma measurements than measurements of air activity, most of the measurements made were of gamma activity in order to cover the greatest possible area. The highest gamma reading recorded was 4 milli-r per hour on the Bailey Bridge near Sellafield station. This reading was mainly due to the activity in the plume. A second van was sent out about 17.00 to the north of the factory and likewise spent most of its effort in measuring gamma activity in order to delimit the contaminated area. These two vans maintained continuous patrols throughout the night and the next day.

86. Some measurements, however, were made of the air activity on 10th October. The highest reading obtained was on the Calder Farm road at 23.00 and the value was about the same as in the Calder Works, i.e. some 10 times greater than ICRP level for continuous lifetime breathing.

87. The Health Physics Manager had to think of three kinds of hazard – gamma radiation to the whole body, inhalation and ingestion. In our opinion he correctly decided that irradiation and inhalation were likely to prove well within acceptable limits for the district and that the main concern should be ingestion.

88. For a time the Health Physics Manager thought that the activity which was escaping from the top of the stack was normal fission products and he therefore had to plan on the assumption that the main risks would come from Iodine and Strontium. By Friday morning he had obtained the first analysis of the milk samples and the first results on the activity collected from the air in the Windscale Works. There was agreement that the activity contained a far greater proportion of Iodine activity than would normal fission products. Having obtained this result, the explanation was clear. Iodine vapour had

come through the filter but the major part of the particulate material had been caught by the filter.

89. The deposition of Iodine 131 on grass which is eaten by cattle is quickly carried to their milk and a major part of the effort of the teams assessing the activity in the district has been given to milk analysis for Iodine.

90. No established tolerance level existed for radioactive Iodine in milk but the Health Physics Manager had in mind a paper ARC/RBC 5 by Dr. Scott-Russell which suggested 0.39 μc per litre as the level beyond which Iodine in milk would be a hazard to infants.

91. The first analysis of milk showed Iodine 131 contents ranging from traces to 0.48 μc per litre, but when the analysis of the Seascale morning milk of Saturday, 12th October was completed at 3 p.m. and showed 0.8 μc per litre the Health Physics Manager advised the Works General Manager that in his opinion the distribution of the milk gathered from the immediate vicinity of the Works should be stopped.

92. There followed several hours of consultation by telephone and by meetings between the medical and health physics experts, in order to establish the limit of radio-iodine content beyond which milk should be taken out of distribution. This was calculated, by reference to probable absorption into the thyroid glands of young children, at 0.1 μc per litre, a figure which was subsequently endorsed both by the Authority's medical consultants and by a meeting specially convened by the Medical Research Council as representing a reasonable assumption.

93. This decision was reached at about 21.00 on 12th October and by arrangements made locally with the Cumberland police and the Milk Marketing Board milk deliveries from 12 milk producers within a two-mile radius of Windscale were prevented that night. Throughout Sunday and Monday milk samples were collected from farms at increasing distances from Windscale. The scientific analysis was shared between the laboratories at Windscale and at Harwell. As the analyses were completed the restriction on distribution had to be extended in successive stages until by Monday morning it covered a coastal strip approximately thirty miles long, ten miles broad at the southern end and six miles broad in the north. To the south it included the Barrow Peninsula and the northern boundary was about six miles north of Windscale.

94. Samples were taken also around the Lancashire Coast, the North Wales Coast, the Isle of Man, and into Yorkshire and the South of Scotland, but there was no necessity to extend the boundaries of restriction.

95. At the time of writing this report, the recorded levels are decreasing throughout the restricted area; but no part of the area is yet sufficiently clear to enable the Authority to recommend to the Ministry of Agriculture, Fisheries and Food relaxation of the restriction.

96. Other possible sources of ingestion hazard were examined, in particular vegetables, eggs, meat and water supplies. None of these was found to be harmful.

97. A thyroid iodine survey has been made and is continuing among the local inhabitants around the works. The situation is similar to that among the people at the works. The highest thyroid activity measured among adults

and children is 0.28 μc in the gland of a child. This level can be compared with the ICRP safe continuous level for adults of 0.1 μc in the gland. The results of the measurements made are under study by the Medical Research Council.

98. Parallel with the measurements of iodine activity are similar measurements on the two Strontium isotopes Sr. 89 and Sr. 90. Enough results have now been obtained to suggest that there is no Strontium hazard arising from the accident. Equally there is no hazard arising from radioactive Caesium.

99. It was represented to us that warning of an emergency ought to have been conveyed to the inhabitants of the area surrounding the Works. Two witnesses reported that high levels of activity had been measured on grass and on clothing in Seascale, and on clothing of people cycling to work along the track from Seascale on the morning of 11th October. The Health Physics Manager was satisfied, from the district measurements already mentioned, that no district radiation or inhalation hazard existed: there was therefore no occasion to issue a district emergency warning, which would have caused unnecessary alarm. The activity measured on the clothing of the two witnesses mentioned above was some 20 times lower than that which would have constituted any hazard in accordance with the standards observed by the Authority and based on the Medical Research Council tolerance levels.

CHAPTER VII. CONCLUSIONS

100. Our conclusions on the three points specifically referred to us are as follows:-

(a) The cause of the accident was as described in Chapter II.
(b) The steps taken to deal with the accident, once it had been discovered, were prompt and efficient and displayed considerable devotion to duty on the part of all concerned.
(c) The measures taken to deal with the consequences of the accident were also adequate. There has been no immediate damage to the health of any of the public or of the workers at Windscale, and it is most unlikely that any harmful effects will develop. No physical damage has been sustained to property other than to Pile No. 1.

101. A major technical defect contributing to the accident was inadequacy of instrumentation for the safe and proper operation of a Wigner release. The uranium thermocouples were correctly positioned for normal operations, but not for Wigner releases; moreover, in the condition of stagnant air which had necessarily been created within the pile, there was no means of detecting the smouldering uranium cartridges which we believe to have been a key event in the development of the accident.

102. The absence of an operating manual for Wigner releases must be regarded as a serious defect. Brief instructions such as the minute reproduced at Appendix C [reproduced on p. 167] are clearly inadequate, particularly for the circumstances prevailing in a second nuclear heating. We were obliged to piece together the other details of pile operation from

Committee minutes and from traditions which did not seem to have any written authority.

103. One minor point should be mentioned. The Pile is a very large structure and during a Wigner release operation it is frequently necessary for the people in charge of the release to climb some 70 feet to the roof in order to inspect the thermocouple readings. There is no lift; its absence must have added considerably to the operational difficulties.

104. The evidence which we received revealed deficiencies and inadequacies of organisation.

105. The division of responsibility between the Operations Branch, the Research and Development Branch of the Industrial Group, other technical branches within the Industrial Group and technical advisers at Harwell is not, in our opinion, clearly defined. Technical changes made by one team within the Authority have not always been known to others who should have been aware of them for the proper discharge of their technical responsibilities and in some cases have not been adequately taken into consideration by those who were aware of them. There is uncertainty as to who is responsible for particular technical decisions, and there is undue reliance on technical direction by committee.

106. The operations staff at Windscale are not well supported in all respects by technical advice. For example, it appears that the records of recent Wigner releases had not been discussed by the Windscale Works Technical Committee. It appears also that changes have been made in operating procedures, with a general tendency to push pile temperatures upwards, without complete realisation of all the technical factors involved. We feel that, since the Windscale Piles were handed over to the operations staff, other demands on the Industrial Group have been so heavy that insufficient technical attention has been available to ensure the safe operation of the Windscale Piles.

107. In our view, one of the lessons of the accident is that the Windscale organisation is not strong enough to carry the heavy responsibilities at present laid upon it. The responsibilities of the Works General Manager include the Calder Hall and Chapelcross reactors as well as Windscale. No one individual should be expected to exercise day to day operational control of so vast an organisation. We understand, moreover, that several of the posts in the Windscale complement under the Works General Manager were unfilled at the time of the accident.

108. The evidence which we received indicated that, when the accident occurred, the several responsibilities of the Chief Safety Officer, the Group Medical Officer and the Windscale Health Physics Manager were not clearly defined. Moreover, it appears to us to be unsatisfactory that tolerance levels in respect of several of the possible hazards should have had to be worked out in haste after the accident had happened.

109. While in the above conclusions we have naturally concentrated on faults and deficiencies, we must pay tribute to the efficient and energetic way in which the accident was dealt with once it was appreciated. Due to the efforts of the Windscale staff, a worse accident was averted.

CHAPTER VIII. RECOMMENDATIONS

110. Conscious of the great public anxiety concerning this accident, we felt we should report as soon as we had been able to consider the technical evidence sufficiently to discharge our terms of reference. We have not, however, in ten days been able to make a full technical assessment of all aspects of this matter. Moreover we are not properly constituted to recommend detailed organisational changes.

111. We therefore recommend:-

(a) That a Technical Evaluation Working Party should be set up within the Authority to make an urgent and thorough study of all the technical information to be derived from the accident. The technical records and reports which have been made available to us are being preserved in order that they may be studied by such a group.

(b) That the Authority should review the organisation of the Industrial Group with regard to the relationship between the operational staff of the Windscale Piles and other technical directorates, and the adequacy of staff both in numbers and quality to cope with the responsibilities laid upon them.

(c) That the responsibilities within the Authority for control of health and safety should be clarified.

(d) That steps should be taken to ensure that maximum permissible levels are laid down for all radioactive substances for exposure for a limited period as well as for continuous exposure. (We recognise that the responsibility for this does not rest solely with the Authority.)

(e) That Pile No. II should not be restarted until its instrumentation is fully adequate for the safe operation of a Wigner release, and until a careful review has been made of the factors involved in the controlled release of Wigner energy.

26th October, 1957

Note on Sources

1. The main bulk of documentary sources is in AEA files and scientific reports (see bibliography) but there are a few important files in the No.10 Downing Street Series (PREM/2156), the Department of Energy and MAFF. Most of the more important documents may be found in many files, and the references given indicate only where they were located by the author.

Public Record Office Documentation

2. Virtually all the relevant AEA documents are now in the PRO, some having been opened early in order to complete the record. They are all in PRO classes prefixed AB, and they include:

(a) Three omnibus files on the accident apparently compiled soon after the event: AB16/2441, 2442 and 2443.
(b) Sir Leonard Owen's comprehensive file AB38/51 (Sir Leonard Owen was Managing Director of the AEA's IG).
(c) The relevant Board papers and minutes (AEA and AEX) for 1957 and 1958: AB16/2704 and 2705.
(d) The Penney Inquiry report, AB86/25, and Inquiry material (77 pieces all in the PRO class AB86).
(e) Scientific reports and memoranda by the IG and the Research Group, in classes AB7 and AB15 respectively.
(f) AEA (London Office) files, in class AB16.
(g) AEA (Risley) files, in classes AB8, AB9, AB19 and AB38.
(h) AEA (Harwell) records, in classes AB6 and AB12.
(i) Windscale files returned by BNFL, in class AB62.
(j) Various Committee minutes and papers: see para. 3 below.

Committees

3. Besides the AEA and AEX (see para. 2(c) above) the Committees of particular interest are:

Fleck Organisation Committee: AB16/2234, 2235 and 2236
Fleck Health and Safety Committee: AB16/2234, 2235, 2236, 2270, 2433, 2666, 2667, 2668, 2669 and 2671
Fleck Technical Evaluation Committee (TEC): AB38/36, AB16/2434, and bound volume AB16/2703
TEC Filter Working Party: AB12/380 and AB16/2707
TEC Graphite Working Party: AB12/331
TEC Wigner Release Working Party: AB12/493
TEC Instrumentation Working Party: AB9/392
TEC Cartridge Working Party: AB9/391
TEC Operations Working Party: AB9/512

Technical Executive Committee (IG): AB9/393, 394 and 395; and AB38/170 and 171
Reactor Safety Research Panel 1955–57: AB12/242
Pile Operating Committee 1947–50: AB12/38 and 179
Pile Operating Committee, Safety Sub-Committee 1956–57: AB12/304
Health Committee 1957–60: AB12/345

Relevant AEA Files

4. A number of useful and relevant AEA files are noted below under various major topics: (a) the Windscale piles (general); (b) uranium fuel elements; (c) AM cartridges; (d) Wigner energy releases in the piles; (e) pre-accident work on graphite; (f) post-accident work on graphite; (g) other relevant research; (h) filters and stack emissions; (i) environmental and site monitoring; (j) the milk ban, and claims for compensation (including non-milk claims); and (k) remedial action. These lists do not pretend to be comprehensive; there are hundreds of items available in the PRO in the AB classes noted in para. 2 above.

(a) *Windscale piles (general)*

AB86/88	Chart of reactor
AB16/2699	History of Windscale by H. G. Davey
AB62/57 and 58	Notes for H. G. Davey's book
AB86/59	Summary of Windscale Technical Committee minutes
AB62/32	Pile Group programme
AB62/34	Pile Group reports
AB62/35	Incidents in Pile No.1
AB62/42	Operational problems of Windscale piles
AB62/72	Bailey papers (general description of piles)

(b) *Uranium fuel elements*
  (i) Oxidation of uranium: AB7/1931 and 2859; AB15/5594; AB86/35, 36, 79 and 84
  (ii) Burst rate: AB7/1962 and 5759; AB86/73, 82 and 86

(c) *AM cartridges*
AB86/33, 38, 46 and 57; and AB62/74

(d) *Wigner energy releases in the Windscale piles*

AB7/1655	Unexpected temperature rise, Pile No.1, September 1952 (IG Report 5006/78)
AB7/2408	Wigner energy release, Pile No.2, August and October 1953
AB7/2575	Wigner energy release, Pile No.2, November 1953
AB86/78 AB86/87 AB9/123	Wigner energy releases in Pile No.1

AB86/27 } AB86/31 } AB86/32 }	Graphs of Wigner release October 1957
AB8/69	Instruction of 14 November 1955 on Wigner releases
AB86/37	Spread of Wigner reaction in last five attempted anneals
AB86/47	Pile programme October 1957
AB62/40 } AB62/41 }	History of Wigner releases in Pile No. 2

(e) *Pre-accident work on graphite*
AB7/2923 and 6136; AB9/391, AB15/5057; and AB86/53, 72 and 77
(f) *Post-accident work on graphite*
AB8/819 and 820; AB9/394 and 517; AB12/331 and 493; AB15/6106; and AB38/107
(g) *Other relevant research*
AB7/6435 and AB86/40
(h) *Filters and stack emissions*
AB12/380; AB15/6106; AB16/2707; AB62/47; and AB86/76
(i) *Environmental and site monitoring and measurements, pre- and post- accident (and meteorological observations)*
AB6/2067, 2068, 2069, 2070, 2071, 2072, 2073; AB15/6033; AB16/2689 and 2691; AB62/33, 53 and 54; AB86/63, 64, 65, 66 and 67
(j) *Milk ban, and compensation claims*
  (i) Milk: AB16/2326, 2431, 2432, 2695 and 2696
  (ii) Compensation claims other than milk: AB16/2371, 2328, 2432 and 2692; and AB8/763
(k) *Remedial action*
AB62/44, 46, 62, 70, 71 and 73

5. Other files of special interest include AB16/2301 on public relations, AB16/2318 on staff relations, and AB7/6143 and AB62/49 on discussions of the accident with US scientists at conferences in December 1957 and January 1958.

Notes and References

(All references are to AEA files in the PRO unless otherwise indicated.)

Introduction

1. Letter Strauss/Penney, 23 February 1960, on file AB6/2071.
2. In 'Statements by the Prime Minister and Mr Churchill on the Atomic Bomb, 6 August 1945', reprinted in M. Gowing, *Independence and Deterrence: Britain and Atomic Energy 1945–52* (Macmillan 1974), vol. 1, pp. 14–18.
3. The programme was announced in February 1955 (Cmd 9389), expanded in April 1957 (Cmnd 132) and stretched out in June 1960 (Cmnd 1083).

1: Britain's Atomic Bomb

1. The pre-war and wartime section of this chapter is based on M. Gowing, *Britain and Atomic Energy 1939–1945* (Macmillan 1964). The pre-war history is recounted in a long introductory chapter by K. E. B. Jay. The post-war section of this chapter is based on M. Gowing, *Independence and Deterrence: Britain and Atomic Energy 1945–1952* (Macmillan 1974). See also M. Gowing, 'Britain, America and the Bomb', in D. Dilks (ed.), *The Retreat from Power* (Macmillan 1981); M. Gowing, 'Nuclear Weapons and the Special Relationship', in William Roger Louis and Hedley Bull (eds), *The Special Relationship – Anglo-American Relations since 1945* (Clarendon Press 1986); M. Gowing, 'How Britain produced the Bomb', *Guardian*, 8 April 1985; L. Arnold, *A Very Special Relationship: British Atomic Weapon Tests in Australia* (HMSO 1987).
2. The memorandum is in two parts; the first part was reproduced in Gowing, *Britain and Atomic Energy 1939–1945*, as Appendix 1. The second part, which was found later elsewhere, was reproduced in Ronald Clark's biography of Sir Henry Tizard. The whole document is now in the PRO (AB1/210).
3. '94' was the new element's atomic number (i.e., the number of protons in the nucleus, which determined the element's chemical identity). '239' was the atomic mass of one of its isotopes, plutonium-239, having in the nucleus 94 protons and 145 neutrons – a total of 239 nucleons.
4. The MAUD report is reproduced in full in Gowing, *Britain and Atomic Energy 1939–1945*, as Appendix 2, pp. 394–436.
5. The Quebec Agreement is reproduced in Gowing, ibid., as Appendix 4, pp. 439–40.
6. The Hyde Park Agreement of September 1944 is reproduced in Gowing, ibid., as Appendix 8, p. 447. The Washington Declaration of November 1945 is reproduced in Gowing, *Independence and Deterrence*, vol. 1, as Appendix 4, pp. 82–4.
7. The second part of this chapter, on the post-war period, is based on Gowing, *Independence and Deterrence*.

2: Windscale's Origins

1. This chapter is based mainly but not exclusively on M. Gowing, *Independence and Deterrence: Britain and Atomic Energy 1945-52* (Macmillan 1974), especially on vol. 2, Chs 18, 19, 21 and 22.
2. See also K. E. B. Jay, *Britain's Atomic Factories* (HMSO 1954), and S. Sinclair, *Windscale - Problems of Civil Construction and Maintenance* (George Newnes 1960).
3. Lecture 27 January 1967 by Sir John Cockcroft, 'Harwell's twenty-first anniversary', published in *Atom*, 125 (March 1967), pp. 51–8. Also meeting at Risley 10 May 1951 and paper WHC No.37 dated 4 July 1951, also notes by Hinton dated 8 May 1951, all on file AB62/47.
4. See Gowing, *Independence and Deterrence*, vol. 2, Chs 23 and 24, and L. Arnold, *A Very Special Relationship: British Atomic Weapon Tests in Australia* (HMSO 1987), Ch. 3.

3: After Hurricane

1. See M. Gowing, *Independence and Deterrence: Britain and Atomic Energy 1945-52* (Macmillan 1974), vol. 1, pp. 421–36, and R. Williams, *The Nuclear Power Decisions* (Croom Helm 1980), pp. 21–5. See also L. Arnold, 'The birth of the UKAEA', *Atom* 225 (July 1975), pp. 94–9; and UKAEA First Annual Report 1954–55.
2. Cmd 8986 (HMSO 1953).
3. Personal communication.
4. See L. Arnold, *A Very Special Relationship: British Atomic Weapon Tests in Australia* (HMSO 1987), p. 76.
5. Gowing, *Independence and Deterrence*, vol. 2, Ch. 19.
6. See J. Simpson, *The Independent Nuclear State* (Macmillan 1986), p. 101; UKAEA Second Annual Report 1955–56.
7. See Arnold, 'The Birth of the UKAEA'.
8. Cmd 9389 (HMSO 1955).
9. Cmnd 132 (HMSO 1957).
10. Cmnd 1083 (HMSO 1960).
    All were twin-reactor Magnox stations: Berkeley and Bradwell (commissioned 1962); Hunterston A (1964); Trawsfynydd, Dungeness A and Hinkley Point A (1965); Sizewell A (1966); Oldbury (1967); and Wylfa (1971). They ranged in size from 250 megawatts (Bradwell) to 840 megawatts (Wylfa).
11. Cmnd 2335 (HMSO 1964). It proposed a programme of 5000 megawatts capacity to be built 1970–75.
12. See Arnold, *A Very Special Relationship*, p. 76.
13. HOC Debates, 1 March 1955, col. 1899.
14. Cmd 9391 (HMSO 1955).
15. UKAEA First Annual Report 1954–55. para. 152.
16. UKAEA First Annual Report 1954–55, para. 155.
17. Letter Hinton/Fleck 26 November 1957 on file AB16/2234.
18. Gowing, *Independence and Deterrence*, vol. 1, p. 229.

19. Ibid., vol. 2, p. 90.
20. Cmnd 338 (HMSO 1957), para. 57.
21. Paper dated 21 October 1957 by P. E. Carter on file AB38-71.
22. Paper P13 dated 27 August 1953 by J. M. Hill on allocation of space for irradiation in the Windscale piles, on file AB62/57.
23. From notes on the pile group by H. G. Davey, in file AB62/57.
24. Gowing, *Independence and Deterrence*, vol. 2, pp. 392–3.
25. Report IG 5006/27, 'The unexpected temperature rise on Pile No. 1 on 30.9.1952', on files AB62/34 and AB7/1655. Meeting at Harwell 28 November 1952 to consider unexpected temperature rise, on files AB62/42 and AB7/1655.
26. Report by J. L. Phillips, SRW 1014 – PGR 33 dated 12 February 1953, 'Wigner energy release in Pile No. 2, 2.1.1953', on file AB62/40.
27. See Technical Note 90 WTSC-R85 (September 1953) to Reactor Safety Research Panel, on file AB7/2923.
28. Letter J. L. Phillips/G. Packman, 2 November 1953, on file AB62/55.
29. Note (?October 1955) by J. L. Phillips, 'Further Incidents in the Life of the Windscale Piles', on file AB62/42.
30. Oral statement by H. G. Davey to Penney Inquiry. A detailed account of the conference was written by P. T. Fletcher and T. M. Fry, 'Note on US/UK Joint Meeting on Reactor Hazards, 12, 13, 14 June 1956 at Chicago'. (Copy in the possession of the author.)
31. Meeting 20 December 1951 including D. D. W. Cole, L. Rotherham, J. C. C. Stewart, J. M. Hill, F. M. Leslie and J. L. Phillips concluded that the piles were running comparatively smoothly but with insufficient background knowledge. There were two main problems: how best to run the piles according to the design conditions, and how far beyond the design conditions they could go (file AB62/42).
32. See Appendix V and notes on pile group by H. G. Davey, in file AB62/57.
33. Two meetings at Windscale 12 June 1952 on pile problems, and paper October 1955 by J. L. Phillips, 'Further Incidents in the Life of the Windscale Piles', on file AB62/42.
34. Letter (October 1955 but undated) from the AEA's radiological consultants to the Chairman with enclosures by F. R. Farmer and A. S. MacLean. They estimated that more than 50kg. of uranium had been oxidised and that altogether 100–200g of irradiated material had been released: see file AB12/207.
35. Note of meeting 27 September 1955, on file AB12/207.
36. Notes of meeting of consultants on 30 September 1955, and supporting papers by McLean and Farmer on 'The release of particulate activity from Windscale works', on file AB12/207. The consultants present were Professor W. V. Mayneord, Sir Ernest Rock Carling, Professor J. S. Mitchell and Dr J. F. Loutit.
37. Windscale paper IGO-R/R6 (IGC-PHC/P13), 'The release of oxide from irradiated uranium during October 1955', on file AB12/207.
38. Howells/Marley, 12 July 1957; AERE (Atomic Energy Research Establishment) D/M 324 dated 30 September 1957; letter, Cockcroft/Himsworth, 30 September 1957; Marley/Cockcroft 1 October 1957; all on file AB6/2352. Also AEX papers: AEX(57)54 dated 2 August 1957;

AEX(57)57 dated 30 August 1957; AEX(57)66 dated 26 September 1957.
39. MRC 57/670, 'Note on Visit to the UKAEA factories at Windscale and Calder Hall on 17 July 1957', on AEA file AB6/2352 and Department of Energy file EG1/96. Letter Keely (MAFF)/Peirson (AEA) dated 31 July 1957 (MAFF reference Z 862), and meeting 2 August 1957 on Department of Energy file AEA 57 Pt II (Annex).
40. The history of pre-accident emissions from Windscale is summarised in a paper by E. H. Underwood for the AEA Chairman, 8 December 1958, on file AB16/2441.
41. Letter Carnochan/Thompson, 14 January 1959 refers to Prime Minister's instructions about secrecy received shortly after a meeting with Plowden on 2 August 1957. Carnochan thought 'the wrong sort of secrecy had been applied in this case': see AB16/2689.
42. For detailed information on the pile stacks, see file AB62/47 *passim*; also the papers of the Fleck (TEC) Filter Working Party, on file AB12/380; also a most useful article by I. A. Mossop, 'Filtration of the gaseous effluent of an air-cooled reactor', *British Chemical Engineering* (June 1960), pp. 420–6.
43. Gowing, *Independence and Deterrence*, vol. 2, p. 395; PEC meeting, 8 May 1952, on file AB9/86. See also Phillips/Tuohy, 12 January 1955, on file AB62/47 for later estimate of cost of filters in terms of production.
44. Notes by H. G. Davey on the Pile Group, in file AB62/57.
45. The complements and strengths of Windscale's professional qualified operations staff at this time (excluding H. G. Davey, the Works General Manager, and T. Tuohy, his deputy) were:

	Complement	Strength	Shortfall
Works Managers	2	1	1
Assistant Works Managers	4	3	1
Engineers	106	92	14
Scientists	222	206	16
Totals	334	302	32

(Organisation chart in file AB38/51).
NB: In 1956 the qualified scientists and engineers at Harwell totalled 1265.

46. Letter Hinton/Davey, dated 26 August 1957 (copy in author's possession).
47. MS letter Davey/Hinton, dated 19 November 1957 (copy in author's possession).

4: The Ninth Anneal

The narrative in this chapter is based on files AB16/2441, 2442 and 2443; the report of the Penney Inquiry; evidence given to the Inquiry; interviews, conversations and correspondence over many years with (in alphabetical order) Mr A. M. Allen, Mr H. J. Dunster, Dr Arthur Chamberlain, the late Mr W. Crone, Professor F. R. Farmer, Mr R. Gausden, Mr V. Goodwin, Sir John Hill, the late Lord Hinton, Mr H. Howells, Mr T. C. Hughes, Mr G. D. Ireland, the late Dr A. S. McLean, Mr R. I. Robertson, Mr K. Saddington and Mr T. Tuohy.

1. One recommendation made after the accident was that more sophisticated meteorological provision was needed at nuclear sites.
2. It reached Belgium late on 11 October, Frankfurt late on 12 October and Norway some time on 15 October. The path of the cloud was followed at the time by means of air sampling and more recently has been calculated by the use of trajectory analysis. Varying accounts of the meteorological conditions are given in Cmnd 302 (HMSO 1957), J. Crabtree 'The travel and diffusion of radioactive material emitted during the Windscale accident: *Journal of the Royal Meteorological Society* 85 (1977); Cmnd 1225, Appendix H (HMSO 1960); H. J. Dunster, H. Howells and W. L. Templeton, 'District surveys following the Windscale accident, 1957. Proceedings of the 2nd UN Conference on Peaceful Uses of Atomic Energy 18 (1958) pp. 296–308; H. M. ApSimon, D. Earnshaw, A. J. H. Goddard and J. Wrigley, 'Trajectory Analysis with particular reference to the 1957 Windscale release' (Imperial College of Science and Technology, Nuclear Power Section, London 1977).

5: Damage Assessment and Damage Control

1. Plowden/PM, 11 and 12 October 1957, on files AB16/2441 and PREM11/2156.
2. PM/Plowden, 13 October 1957, on files AB16/2441 and PREM11/2156.
3. Plowden/PM, 14 October 1957, *ibid.*
4. MAFF and MHLG (then the department responsible for water supplies) were the departments statutorily concerned with any radioactive releases from Authority sites. The MRC was not represented and apologies were later made to it. Meeting 14 October 1957, on files AB16/2441, AB6/2352 and AB38/51.
5. MAFF/AEA correspondence November 1956–January 1957, on file AB6/2067.
6. 'An Assessment of hazards resulting from the ingestion of fallout by grazing animals', R. Scott Russell, AERE ARC/RBC 5 (1956).
7. Meeting 15 October 1957, on files AB16/2441, AB6/2352, AB38/51 and AB38/170. The consultants at the meeting were Sir Ernest Rock Carling, Dr J. F. Loutit, Professor W. V. Mayneord and Dr R. Scott Russell.
8. Note by J. Wyndham 16 October 1957, Plowden/PM 17 October 1957, on files AB16/2441 and 2442.
9. There are two separate accounts of this meeting (a preliminary one by Himsworth and the official minutes MRC57/795), on files AB6/2072, AB6/2352 and AB16/2441.
10. The first report on 'Hazards to Man of Nuclear and Allied Radiations' was Cmd 9780 (HMSO June 1956) and the second was Cmnd 1225 (HMSO December 1960).
11. The Authority's three groups (up to the end of 1957) were the Research

Group, the Industrial Group, and the Weapons Group. The Authority Members for Research, for Engineering and Production, and for Weapons Research and Development, were heads of the three Groups as well as being Board Members with a corporate responsibility.

12. The responsible Minister at this time (April 1957–November 1959) was the Prime Minister himself.

    The Lord President of the Council was the Minister responsible for, *inter alia*, atomic energy January 1954–April 1957.
13. The history of these procedures will be found in files AB19/1 and AB16/207; IGC/PEC/223 in PEC, vol. 3 (April 1956–May 1957); and AEA Board papers AEA(56)136 and AEA(57)71.
14. Allen/Bishop 11 October 1957 on files AB16/2441 and PREM11/2156.
15. Statement to press 4.30 p.m. 11 October 1957 on file AB16/2442.
16. PM/Plowden. PM521/57 of 13 October 1957, on files AB16/2441, AB38/51 and PREM11/2156.
17. Highton/Allen, 14 October 1957, on file AB16/2441.
18. Plowden/PM, 14 October 1957, on files AB16/2441, AB38/51 and PREM11/2156.
19. Statement to press, 15 October 1957, on file AB16/2442.
20. Private communication.
21. Papers and minutes of the full Board (AEA) and the Executive (AEX) relating to the Windscale accident are in the PRO, AB16/2704(1957) and 2705(1958). The AEX consisted of the full-time Members of the Board.
22. Telephone message Peirson/Plowden, 17 October 1957, on file AB16/2442.
23. Note, 23 October 1957 by Allen, on file AB16/2441.
24. Telegram, 18 October 1957, Allen/Gaunt, on file AB16/2441.
25. Letter, 22 October 1957, from Frank Anderson, MP, to Whitehaven News (and other local papers), on file AB16/2441.
26. An undated summary of press reactions (to 23 October 1957) is on files AB16/2442 and AB16/2301.
27. Correspondence Hailsham/Quirk, 21–25 October 1957, on Department of Energy file EG1/96.
28. Hampton/Peirson, 21 October 1957, on file AB16/2442; Kronberger/Rotherham *et al.*, on file AB38/51. Also TX(57)1 dated 16 October 1957 on file AB9/393.
29. TX minutes and papers are on file AB9/393.
30. The Windscale Pile Specification Committee on file AB9/393.
31. 'The Windscale accident – notes for the Lord President', 25 October 1957, on file AB16/2441.
32. A set of press releases is enclosed in file AB16/2442. See also file AB16/2301.
33. Mitchell/Owen, 26 October 1957, on file AB38/51.
34. For an interesting account of Davey's highly personal and successful local public relations policy, see 'A pioneer of atomic energy public relations', *Atom*, 70 (August 1962), pp. 165–73.
35. From note by Hill, 16 October 1957, on file AB38/51: 'The meeting with the miners went very well. The miners' representative told the Press they were perfectly satisfied.' From note by Gillams, 31 October 1957, *ibid*: 'A Meeting of Cumberland farmers last week ended with a vote of confidence

in Mr Davey's statements... A Labour member of the Cumberland County Council said the people of Cumberland had every confidence in the Authority and he personally would be only too pleased if health conditions in the mines and elsewhere were as good.'

36. Letter 16 November 1957 from Dr J. N. Dobson (Medical Officer of Health, Ennerdale Rural District Council) to *The Lancet*, on file AB16/2441.
37. About 676 000 gallons. See files AB16/2326 and AB16/2696.
38. See file AB6/2072 *passim*, and MAFF files NE75 and NE75A, especially documents ARC97/58, ARC337A/58, ARC338/58, ARC338A/58, ARC798A/58, ARC800/58.
39. The Joint ARC/MRC/DC Committee on Biological (Non-Medical) Problems of Nuclear Physics.
40. An interesting article from a local newspaper describing local attitudes to Windscale and reactions to the accident is quoted at length in 'A pioneer of atomic energy public relations'.
41. At the time of the Three Mile Island accident which began on 28 March 1979, 60 per cent of the population within 5 miles evacuated the area (some 21 000 persons). Within a radius of 15 miles, 144 000 persons (or about 39 per cent of the population) evacuated, mostly by their own decision, not because they had been formally advised or ordered to do so: Thomas H. Moss and David L. Sills, *The Three Mile Island Nuclear Accident: Lessons and Implications*, Annals of the New York Academy of Sciences, vol. 365 (1981), pp. 148–9.
42. See Fleck Committee paper FC/T8 of 13 December 1957 (Submission from the Clerk to the Cumberland County Council), in bound volume AB16/2703. The Clerk argued that the Authority was in breach of Section 5(3) of the Atomic Energy Authority Act 1954; the AEA Legal Adviser stated that Section 5(3) did not give the AEA the duty of preventing the emission of ionising radiations but the significantly different duty of securing that no ionising radiations from their establishments caused any hurt or damage (see Chapter 7, note 33, below).
43. 'District Survey Programmes 15 October 1957, 10.15 am', on files AB16/2441 and AB38/51; 'Note on radioactive contamination in NW England following the Windscale accident', by H. J. Dunster, on files AB6/2068 and AB6/2352; Dunster, Howells and Templeton, 'District surveys following the Windscale accident, October 1957'. See also file AB16/2691.
44. F. B. Ellis, H. Howells and W. L. Templeton, 'Deposition of strontium 89 and 90 on agricultural land and its entry into milk after the reactor accident at Windscale', AERE AHSB(RP)R2; and file AB16/2689.
45. The work is described in a series of reports on file AB38/51: Hill/Owen, 15 October 1957, 25 October 1957, 26 October 1957, 29 October 1957; McFarlan/Fletcher, 1 November 1957.
46. Press release, 21 October 1957, on file AB16/2442.
47. McFarlan/Fletcher/Ross, 1 November 1957, on file AB38/51.
48. Gausden report, 29 October 1957, on Inspection of Core, on file AB16/2442.
49. Message Plowden/Penney, 17 October 1957, on file AB16/2442; AEA(57)18th meeting, 17 October 1957, on AB16/2704.

50. IG Report, T. N. Rutherford, E. Fanthorpe and P. B. Woods, 'The Windscale piles – situation April 1961' (May 1961), AB62/71.
51. Hill/Owen, 15 October 1957 and 26 October 1957, on file AB38/51.
52. AB6/2079; IGR-TM/W213, 'Windscale Pile No.1 incident – effect on the irradiation programme', and AB7/5819.

6: The Penney Inquiry and the First White Paper

1. *Atom*, 225 (July 1975), p. 100.
2. The report is in the PRO: AB86/25. The audiotapes AB86/1–16 and transcripts AB86/17–24 are still closed.
3. AEX(57)19th meeting, 28 October 1957, on file AB16/2704.
4. AEA(57)95, dated 29 October 1957, on file AB16/2704.
5. Allen/No.10 Downing Street, 28 October 1957, on file AB16/2441.
6. Annotated Penney report on file PREM11/2156.
7. PM/Hailsham, 29 October 1957, on file PREM11/2156.
8. Hailsham/Himsworth, 29 October 1957, on files AB16/2441 and PREM11/2156.
9. Numerous letters, dated 29 October 1957 and 30 October 1957, from A. M. Allen on file AB16/2442. The file copy of the letter to the Private Secretary to the Paymaster-General is annotated: 'I warned Wells the following day that this was no longer so and the document was withdrawn on the Friday'.
10. HOC Debates, 29 October 1957, cols 32–36.
11. Brundrett/Penney, 29 October 1957, on file AB86/25.
12. AEA(57)19th meeting, 30 October 1957, on file AB16/2704.
13. AEA(57)95, dated 29 October 1957, on file AB16/2704.
14. Bishop/Michaels, 31 October 1957, on file AB16/2441.
15. Letters from Allen to recipients of Penney report, 1 November 1957, on file AB16/2442.
16. Citrine/Plowden, 15 November 1957, and Plowden/Citrine, 20 November 1957, on files AB38/51 and AB16/2442.
17. Plowden wrote personally to Hinton on 5 November 1957 (file AB16/2442):

> I would have liked to have had a talk with you before the Prime Minister makes his statement about the Windscale accident but as you are on holiday this is not possible. It is not possible in a letter to tell you what the report will contain but there is one thing I would like you to learn from me and not from the newspapers. It is that the Committee of Inquiry attribute a large part of the blame for the accident to the deficiencies of the Operations Directorate of the Industrial Group. They do however make clear that when the accident was known to have happened the staff at Windscale acted with great resource and devotion to duty.

Plowden wrote to Hinton enclosing a copy of Cmnd 302 on the day of publication, 8 November 1957, on file PREM11/2156.
18. Mills/PM 1 November 1957 on files AB16/2441 and PREM 11/2156.
19. Alistair Horne, *Macmillan 1957–1986*, vol. II (Macmillan 1989) p. 54.

20. AEA(57)20th meeting, 4 November 1957, on file AB16/2704.
21. Unnumbered telegram from Washington Embassy Gaunt/Allen, 3 November 1957, on file AB16/2441.
22. See L. Arnold, *A Very Special Relationship: British Atomic Weapon Tests in Australia* (HMSO 1987), *passim*.
23. Horne, *Macmillan*, p. 53.
24. See J. Simpson, *The Independent Nuclear State* (Macmillan, 1983), p. 128, and Arnold, *A Very Special Relationship*.
25. After the atomic project was removed from a government department (the Ministry of Supply) with the creation of the AEA in 1954, a very small department, the Atomic Energy Office, was set up to handle atomic policy in Whitehall. This 'bus-load' of civil servants was responsible at first to the Lord President of the Council, Lord Salisbury (1954–57), then to the Prime Minister, Mr Macmillan (1957–59), then to Lord Hailsham as Lord Privy Seal and Minister for Science (until July 1960) and then Lord President of the Council and Minister for Science (until 1964).
26. 'Accident at Windscale Pile No.1 on 10 October 1957' (Cmnd 302, HMSO November 1957), p. 3.
27. Ibid., pp. 3–4.
28. Ibid., p. 6, para. 15.
29. Ibid., p. 21, paras 2 and 6.
30. Personal communication.
31. Peirson/Perrott 22 November 1957 on file AB16/2441.
32. Personal communication.
33. Sir Harold Himsworth, Sir Ernest Rock Carling, Professor A. Bradford Hall, Dr J. F. Loutit, Professor W. V. Mayneord, Professor J. S. Mitchell, Dr E. E. Pochin, Professor B. W. Windeyer.
34. Cmnd 302, p. 20, para. 29(ii). The MRC Committee also referred in the Annex to the hazard of radiostrontium distributed over the world as a result of nuclear weapon tests, and the possible contributions to the total contamination by emissions from the Windscale plant prior to the accident.
35. Cmnd 302, pp. 22–3.
36. Paper on Wigner energy 12 April 1957 by Cottrell and Thompson on files AB6/1966 And AB6/1971.
37. Undated memo (? May 1958) on Wigner energy on file AB6/1966.
38. M. Gowing, *Independence and Deterrence: Britain and Atomic Energy 1945–52* (Macmillan 1974), vol. 2, p. 104.
39. Tech. Note 90, WTSC/R85 by J. C. Bell, September 1953 on file AB7/2923; summary note by Mummery, 23 October 1957, on graphite oxidation, memo Dunworth/Schonland, 14 November 1957, and memo Schonland/Penney, 18 January 1960, on file AB6/1966.
40. Kavanagh/Stewart (undated, late 1957) with enclosures, on file AB12/331.
41. H. G. Davey in evidence to Penney Inquiry.
42. Minutes of co-ordination meeting, chaired by Cockcroft, at Capenhurst on 22 February 1955, on file AB6/1966.
43. Minutes of Technical Executive Committee, TX(57)M3 11 November 1957, on file AB9/393.
44. Schonland/Cockcroft, 15 November 1957 (on Wigner energy project) and

Cockcroft/Owen, 18 November 1957, also undated paper by Cottrell, on file AB6/1966.
45. Schonland/Cockcroft, 15 November 1957, on file AB6/1966.
46. Penney/Cottrell, 18 November 1957, on file AB6/1966.
47. Cottrell/Cockcroft, 23 May 1958, on file AB6/1966. Progress reports in WEP series on Wigner energy project by Greenough, 26 November 1958, 19 January 1959, 16 April 1959, 15 May 1959, 12 August 1959, and final report 1 June 1960, on AB6/1966. Also reports dated 14 May 1958, 4 June 1958, 2 October 1958, 21 October 1958, 28 October 1958 and 9 December 1958, on file AB9/517.
48. *Atom*, 404 (June 1990), p. 2.
49. Cmnd 302, p. 21, para. 2.
50. PM/Fleck, 15 November 1957, on file PREM11/2156.
51. Bishop/PM, 5 November 1957, on PREM11/2156.
52. CC(57)78th conclusions, Min.4, on PREM11/2156.
53. An undated summary of press coverage is on files AB16/2442 and AB38/51.
54. The comments in *The Economist* of 19 October 1957 had prompted the Lord President's Office to raise the question of the form of the accident inquiry (see Chapter 5).
55. PM/Plowden, 11 November 1957, on file PREM11/2156.
56. Plowden/PM, 12 November 1957, on file PREM11/2156.
57. See note 53 above.
58. Mayne/Plowden, 15 November 1957, and Mayne/Perrott, 2 December 1957, on file AB16/2318; also Annex A to Note of meeting between AEA Official Side and IPCS at Windscale, 21 February 1958, on file AB38/51.
59. *The Economist*, 19 October 1957.
60. Cable Libby (Acting Chairman, USAEC) to Plowden, 12 October 1957, on file AB38/51.
61. Peirson/Strath, 8 November 1957, on file AB16/2442.
62. IGT-TM-020 by N. L. Franklin and W. W. Harpur, 'Notes on discussions with USAEC Mission on Windscale incident 2–4 December 1957', on file AB38/51. Also AEX(57)21st meeting, 14 November 1957, and AEX(57)22nd meeting, 28 November 1957, on file AB16/2705.
63. Undated (late 1957) memo Kavanagh/Stewart on classified conference with the USAEC on graphite, on file AB9/394.
64. Cockcroft/McLean, 2 November 1957, and notes by Marley, 17 December 1957, on file AB38/51; note R MS/SRI, on AB62/49.
65. Cable Rome Embassy/FO, 12 October 1957, on file AB16/2301.
66. AEA Overseas Relations Committee papers ORC(57)68 of 16 December 1957 and ORC(58)1st meeting, 29 November 1957, in AEA archives.
67. Cockcroft/Strath, 24 December 1957, on file AB38/51; also ORC(57)68 dated 16 December 1957, ORC(58)1 dated 4 March 1958, ORC(57)10th meeting, 29 November 1957, ORC(57)1st meeting, 3 January 1958, ORC(58)2nd meeting, 7 February 1958, ORC(58)3rd meeting, 12 March 1958 (AEA archives).
68. N. G. Stewart, R. N. Crooks and E. M. R. Fisher, 'Measurements of the radioactivity of the cloud from the accident at Windscale – data submitted to the IGY', Harwell UKAEA, AERE-M857 (April 1961).

7: Three More White Papers

1. Sir Alexander Fleck, KBE, FRS – later Lord Fleck (1961) – held senior posts in ICI from 1937 and became Chairman of the company in 1953. He was, *inter alia*, at various times President of the British Association for the Advancement of Science, Treasurer and Vice-President of the Royal Society and President of the Royal Institution.
2. C. F. Kearton – later Lord Kearton (1970), FRS (1961) – worked in ICI 1933–40, and then in atomic energy in both the UK and USA (1940–45). Hinton tried unsuccessfully to persuade him to join the post-war British atomic energy project as his deputy: see M. Gowing, *Independence and Deterrence: Britain and Atomic Energy 1945–52*, (Macmillan 1974) vol. 2, p. 24). He joined Courtaulds in 1946 and later became Chairman. He was a part-time Member of the AEA 1955–81.
3. Cmnd 338 (HMSO 1957), pp. 15–19, paras 35–52.
4. Ibid., p. 16 para. 39.
5. Ibid., p. 7.
6. Aide Mémoire (undated) for Sir Leonard Owen for discussions with Sir Alexander Fleck, on file AB9/869.
7. Cmnd 338, p. 19, para. 57.
8. Ibid., p. 23, para. 75.
9. Ibid., p. 27, para. 98.
10. Ibid., p. 20, para. 62.
11. Ibid., p. 14, paras 28–29.
12. Ibid., p. 21, paras 63 and 68.
13. Ibid., p. 10, para. 8 and p. 22, para. 74.
14. Ibid., p. 11, para. 12 and p. 23, para. 76.
15. Ibid., p. 16, para. 41; p. 20, para. 58; p. 23, paras 79–80; p. 24, paras 81 and 82; p. 25, paras 88–90.
16. Ibid., p. 24, paras 83–85.
17. Ibid., pp. 27–30, paras 96–108.
18. Ibid., p. 32, paras 114–21.
19. Ibid., p. 32, para. 119.
20. See undated (late 1957) note on IG organisation by Sir Leonard Owen on file AB9/869.
21. MS annotations by Owen on a memorandum D. S. Mitchell/Owen 1 January 1958 on file AB9/869.
22. AEA(58)1st meeting, 9 January 1958, on AB16/2705.
23. Note of meeting Cook/Owen/Perrott *et al.*, 5 March 1958, on file AB9/869.
24. Sir William Cook, FRS, had been Chief of the Royal Naval Scientific Service before becoming Deputy Director of AWRE (Atomic Weapons Research Establishment, 1945–58). He was appointed Authority Member for Engineering and Production in 1958; then (because of successive reorganisations) Member for Development and Engineering in 1959, and Member for Reactors 1961–64.
25. UKAEA Fifth Annual Report 1958/1959, para. 302.
26. UKAEA Annual Reports for 1956/1957 to 1961/1962.
27. UKAEA Fourth Annual Report 1957/1958, paras 76–78.
28. UKAEA Fifth Annual Report 1958/1959, para. 6.

Notes and References

29. Paper by Sir Leonard Owen (copy in the author's possession).
30. UKAEA Fifth Annual Report 1958/1959, para. 17.
31. He was 'sorry to see that the terms of reference of the Health and Safety Committee had been limited to the arrangements of the Authority': Fleck Health and Safety Committee paper FC/0610, 25 November 1957, on file AB16/2670.
32. AEX(57)71, 24 October 1957, on AB16/2704.
33. The Atomic Energy Authority Act 1954 said:

    It shall be the duty of the Authority to secure that no ionising radiations from anything on any premises occupied by them, or any waste discharged (in whatever form) on or from any premises occupied by them, cause any hurt to any person or any damage to any property, whether he or it is on any such premises or elsewhere.

34. Authority General Notice 2/57, attached as Annex 2 to FC/H&SI, 13 November 1957, on file AB16/2666. Full details of the organisation are in AEA(56)127, 22 November 1956.
35. Cmnd 342 (HMSO 1958), p. 13, para. 47 and p. 15, paras 55–56.
36. The complex story of health and safety organisation in the AEA up to 1960 is summarised in an AHSB paper on the subject, dated 22 April 1960, on file AB16/3646.
37. 'Safeguards' did not then have its present meaning, of the IAEA's system of control of plant and special materials to prevent nuclear weapons proliferation (see Glossary).
38. Cmnd 342, pp. 22–25, paras 93–99, 101 and 102X.
39. The Radioactive Substances (Administrative) Committee which developed out of a Cabinet Committee GEN473. It later (1958) became the Interdepartmental Committee on Atomic Health and Safety. This Committee discussed plans for a licensing system for reactors at meetings on 4 January 1957 and 17 May 1957. The subject was taken up again early in 1958, after the Windscale accident.
40. *Training in Radiological Health and Safety* (HMSO, February 1960).
41. See *The Work of the NRPB 1977/80 and a review of the first 10 years* (NRPB 1981).
42. AEA(58)3, 2 January 1958, on file AB16/2705.
43. TEC/FWP – Filter Working Party papers, on files AB12/380 and AB16/270.
44. FGWP – Graphite Working Party papers, on file AB12/331.
45. See draft report of Instrumentation Working Party, on files AB7/6434 and AB9/392.
46. See papers of Cartridge Working Party, on files AB9/391 and AB9/392.
47. File AB62/74.
48. Strath/Chairman, 18 March 1958, on file AB16/2437.
49. Cmnd 471 (HMSO 1958), pp. 18–19, paras 66–68 and 69(iv).
50. Ibid., p. 18, para. 65.
51. Ibid., pp. 7–9, paras 11–22 and pp. 12–19, paras 38–64 and 69(i)–(iv).
52. Ibid., p. 11, para. 29 and p. 11, para. 31.
53. Ibid., p. 10, para. 27 and p. 19, para. 69(v)(c).

54. AEA(58)62 and AEA(58)11th meeting, 26 June 1958, on file AB16/2705.
55. Strath/PM, 11 July 1958, on PREM11/2156.
56. AEA(58)44, 25 April 1958, and AEA(58)7th meeting, 1 May 1958, on file AB16/2705.
57. Press cutting, 16 July 1958, on PREM11/2156.
58. This voluminous exchange in mainly on file AB16/2318 and partly also on files AB38/51 and AB16/2698. It includes some sixteen letters and five detailed memoranda from the IPCS, eight letters from AEA officials, and one from No.10 Downing Street; and records of IPCS/AEA meetings on 27 November 1957, 13 December 1957 and 21 February 1958.
59. 'Notes on Cmnd 302', 15 November 1957 on file AB16/2318.
60. Mayne (IPCS)/PM, 27 November 1957, and Bishop (No.10)/Mayne, 5 December 1957, on file AB16/2318.
61. Record of meeting, 21 February 1958, on files AB16/2318 and AB38/51. Agreed statement, 11 March 1958, on file AB16/2318, reported in *The Times* and the *Daily Express*, 12 March 1958.
62. FC/T12: note by IPCS for Fleck TEC, 25 February 1958, in Fleck Committee bound volume AB16/2703.
63. Searby/Watson, 12 March 1958, on files AB16/2698 and AB16/2318.
64. Perrott/Tonkin, 7 March 1958, on file AB16/2698.
65. FC/T18 correspondence with IPCS in Fleck Committee bound volume AB16/2703.
66. AEA(58)93 October 1958 and AEA(58)17th meeting, 16 October 1958, on file AB16/2705; AB8/819 and AB8/820, *passim*.
67. Plowden/PM, 17 October 1958, Bishop/PM, 20 October 1958, Bishop/Rawlinson, 21 October 1958, on PREM11/2156.
68. AEA(59)34, 10 April 1959, on file AB16/2705.

8: Causes: An Accident Waiting to Happen

1. Cmnd 302 (HMSO November 1957), pp. 3–4.
2. Ibid., p. 5, paras 2 and 3.
3. Ibid., p. 7, para. 16.
4. Ibid., p. 9, paras 33 and 35.
5. Ibid., p. 9, para. 37.
6. Original Penney report (para. 110) on file AB86/25.
7. Fleck Committee FC/T 4th meeting, 19 March 1958, in bound volume AB16/2703.
8. The author inherited this copy by chance in 1959.
9. Technical Executive Committee meeting, 11 November 1957, TX(57)M3, on file AB9/393.
10. J. S. Nairn, 29 October 1957, on file AB7/6435.
11. Memo dated 29 November 1957, on file AB6/1972.
12. Memo dated 17 January 1958, on file AB38/51 and numerous other UKAEA files.
13. See file AB9/391.
14. AEA(58)93, October 1958, and AEA(58)17th meeting, 16 October 1958.
15. This view is confirmed by a scientist who was involved in graphite

monitoring during the year before the fire. After reading the book in draft he wrote:

> I firmly believe that graphite oxidation was the cause of the fire starting, the air reactivity rate having increased greatly as a result of irradiation and contamination by water and airborne materials . . . It is well known that virgin graphite is very difficult indeed to ignite and I doubt whether the deterioration due to radiation and contamination was appreciated at that time. This, of course, is in direct contrast to the present situation where the relevant parameters are regularly monitored and proper allowance made in all fault studies.

16. Memorandum of 17 January 1958; see note 12 above.

9: Appraisals and Reappraisals

1. The units current at the time of the accident are retained because they occur throughout all the relevant documents. They are:

curie – a unit of radioactivity.
roentgen (abbreviated as 'r') – a unit of radiation, in practice equivalent to the rad
rad – a unit of absorbed dose of radiation.
rem – a unit of radiation dose which takes account of the relative biological effectiveness (RBE) of different types of radiation (alpha and beta particles, neutrons, and gamma and X-rays). RBE is now called Q (Quality Factor).

The international units now in use are:

becquerel – 37×10^9 becquerels = 1 curie
gray – 1 gray = 100 rads
sievert – 1 sievert = 100 rem

(See Glossary.)

2. Meeting, 15 October 1957, on files AB16/2441, AB6/2452, AB38/51 and AB38/170.
3. At the meeting on 15 October 1957 the 0.1 microcurie/litre limit for iodine-131 in milk was said to be based on a maximum dose of 10 rads to the child thyroid, but a figure of 20 rads appears in the later account by H. J. Dunster, H. Howells and W. L. Templeton, 'District surveys following the Windscale accident, October 1957'. *Proceedings of the 2nd UN Conference on Peaceful Uses of Atomic Energy* , 18 (Geneva 1958), pp. 296–308.
4. MRC 57/795 on files AB16/2441, AB6/2352 and AB6/2072.
5. Full text of Penney report is on files AB86/25 and PREM11/2156 (see Appendix XI).
6. 'Accident at Windscale No.1 Pile on 10 October 1957' (Cmnd 302), Annex III.

7. Dunster, Howells and Templeton, 'District Surveys following the Windscale accident', p. 296.
8. Recommendations of the ICRP in the *British Journal of Radiology*, Supplement 6 (1955), pp. 7 and 15.
9. *A Second Report to the Medical Research Council on The Hazards to Man of Nuclear and Allied Radiations* (Cmnd 1225, HMSO December 1960).
10. *The Hazards to Man of Nuclear and Allied Radiations – a Report to the Medical Research Council* (Cmd 9780, HMSO June 1956).
11. 'Maximum permissible dietary contamination after the accidental release of radioactive material from a nuclear reactor', *British Medical Journal*, i (1959), 967–9. Reprinted in Cmnd 1225 as Appendix J.
12. J. F. Loutit, W. G. Marley and R. S. Russell, 'The Nuclear Reactor Accident at Windscale – October 1957: Environmental Aspects', published in Cmnd 1225 as Appendix H.
13. See, for example, R. Russell Jones and R. Southwood (eds), *Radiation and Health: The Biological Effects of Low Level Exposure to Ionizing Radiation* (Wiley 1987), p. 141.
14. N. G. Stewart and R. N. Crooks, 'Long-range travel of radioactive cloud from the accident at Windscale', *Nature, Lond.*, LXXXII, 4636 (September 6, 1958) pp. 627–8.
15. Joh. Blok, R. H. Dekker and C. J. H. Lock 'Increased atmospheric radioactivity in the Netherlands after the Windscale accident', *Applied Scientific Research*, 7 (1958), pp. 150–52.
16. Marley and Stewart/Cockcroft, 6 November 1957, on file AB6/2352.
17. Recommendations of the ICRP, adopted 9 September 1958 (Pergamon Press, 1959), p. 9.
18. J. R. Beattie, 'An assessment of environmental hazard from fission product releases', AEA report AHSB(S)R64 (1963), file AB7/11617.
19. R. H. Clarke, 'An analysis of the 1957 Windscale accident using the WEERIE code', *Annals of Nuclear Science and Engineering*, 1 (1974), pp. 73–82.
20. K. F. Baverstock and J. Vennart, 'Emergency reference levels for reactor accidents: a re-examination of the Windscale reactor accident', *Health Physics*, Vol. 30 (1976), pp. 339–44.
21. A. C. Chamberlain, 'Emission of fission products and other activities during the accident to Windscale Pile No.1 in October 1957' (AEA report AERE-M3194, 1981; declassified 1983).
22. P. J. Taylor, *The Windscale Fire, October 1957*, Research Report RR-7 (PERG July 1981).
23. M. J. Crick and G. S. Linsley, 'An assessment of the radiological impact of the Windscale reactor fire, October 1957' (NRPB R135, May 1982).
24. M. J. Crick and G. S. Linsley, 'An assessment of the radiological impact of the Windscale reactor fire, October 1957' (NRPB R135, Addendum September 1983).
25. The estimate of the collective dose (or, more precisely, collective dose equivalent commitment) was increased from 1.2×10^5 person-rem to 2.0×10^5 person-rem.
26. Sir Douglas Black, *Investigation of the Possible Increased Incidence of Cancer in*

West Cumbria: Report of the Independent Advisory Group (HMSO 1984).
27. M. Bobrow, The Implications of the New Data on the Releases from Sellafield in the 1950s for the Conclusions of the Report on the Investigation of the Possible Increased Incidence of Cancer in West Cumbria: First Report of the Committee on Medical Aspects of Radiation in the Environment, COMARE (HMSO 1986).
28. D. Jakeman, 'Notes on the level of radioactive contamination in the Sellafield area arising from discharges in the early 1950s', AEA report AEEW-R2104 (1986).
29. A. C. Chamberlain, 'Environmental impact of particles emitted from Windscale piles 1954–7', AERE R 12163, April 1986 (information submitted to COMARE). The Science of the Total Environment, 63 (1987), pp. 139–60, Elsevier Science Publishers BV Amsterdam.
30. J. W. Stather, R. H. Clarke and K. P. Duncan, 'The risk of childhood leukaemia near nuclear installations', NRPB-R215 (1988).
31. M. J. Gardner, A. J. Hall, S. Downes and J. D. Terrell, 'Result of case-control study of leukaemia and lymphoma among young people near Sellafield nuclear plant in West Cumbria', British Medical Journal, 300 (17 February 1990), pp. 423–33.
32. V. R. Goodwin 'Radiological and medical consequences of the Windscale fire October 1957' (unpublished paper 23 December 1990)
33. J. H. Gittus (ed), 'The Chernobyl accident and its consequences', AEA NOR 4200, (second edition) April 1988. A very detailed and up-to-date account of the accident has recently been published by Z. Medvedev, The Legacy of Chernobyl (Basil Blackwell 1990).
34. 'National Radiological Protection Board guidance on risk estimates and dose limits', NRPB-GS9, November 1987.
35. 'Report to the General Assembly of the United Nations Scientific Committee on the Effects of Atomic Radiation' (Vienna: United Nations 1988).
36. R. H. Clarke, 'The 1957 Windscale accident revisited', Paper presented at REAC/TS International Conference on the Medical Basis for Radiation Accident Preparedness, at Oak Ridge, Tennessee, 20–22 October 1988.

10: Postscript

1. MS comments on memorandum Mitchell/Owen, 1 January 1958, on file AB9/869.
2. FC/010 25 November 1957 and Hinton/Fleck, 26 November 1957, on file AB16/2234.
3. 'Safeguards' here meant engineering safety of plant, as distinct from radiological protection of persons. Later the term 'safeguards' was applied to the control of nuclear materials to prevent weapons proliferation.
    The Safeguards Division of AHSB became the SRD, and the RPD was absorbed into the NRPB when it was set up in 1970.
4. The Inspectorate was set up by the Nuclear Installations (Licensing and Insurance) Act 1959. This Act was followed by the Nuclear Installations Acts of 1965 and 1969.
5. The Fleck report on organisation (Cmnd 338) praised the 'extraordinary degree of vitality and efficiency with which the Industrial Group [had]

carried out the responsibilities laid on them, accepting individually and collectively a programme of work which kept them continually under the most severe strain' (paras 39 and 50).
6. See Schonland/Cockcroft, 8 January 1958, on file AB6/1971.
7. M. Gowing, *Independence and Deterrence: Britain and Atomic Energy 1945–1952* (Macmillan 1974), vol. 1, p. 229.
8. The Labour MP, Roy Mason, raised the question of the future of the Windscale piles on 6 November 1958 in the House of Commons (HOC Debates, cols 1106–7), and wrote to the Prime Minister on 12 December 1958 asking if any consideration had been given to their dismantling and disposal; whether the governing factor was their mammoth size and solidity or their intense radioactivity; whether ease of dismantling was taken into account in designing atomic power stations which might have a life of only 20 years; what would be the costs; and whether underground siting was feasible. Dismantling, it was said, could not begin for, say, 10 years, would take 2–3 years, the cost would be high, and new provision would be needed for radioactive waste disposal.

Bibliography

Books

Lorna *Arnold, A Very Special Relationship: British Atomic Weapon Tests in Australia* (HMSO 1987).
John C. *Chicken, Nuclear Power Hazard Control Policy* (Pergamon Press 1982).
Margaret *Gowing, Britain and Atomic Energy 1939–1945* (Macmillan 1964).
Margaret *Gowing, Independence and Deterrence: Britain and Atomic Energy 1945–1952*, 2 vols (Macmillan 1974).
Tony *Hall, Nuclear Politics* (Penguin 1986).
Alistair *Horne, Macmillan 1957–1986*, vol. II of the official biography (Macmillan 1989).
Barrie *Lambert, How Safe is Safe? Radiation Controversies Explained* (Unwin Paperbacks, 1990).
Alwyn *McKay, The Making of the Atomic Age* (Oxford University Press 1984).
Zhores *Medvedev, The Legacy of Chernobyl* (Basil Blackwell 1990).
Thomas H. *Moss* and David L. *Sills* (eds), *The Three Mile Island Nuclear Accident: Lessons and Implications* Annals of New York Academy of Sciences, vol. 365 (1981).
Stan *Openshaw, Nuclear Power: Siting and Safety* (Routledge & Kegan Paul 1986).
Walter *Patterson, Nuclear Power* (Penguin 1976, 1983).
Edward *Pochin, Nuclear Radiation: Risks and Benefits* (Oxford University Press 1983).
R. F. *Pocock, Nuclear Power: Its Development in the United Kingdom* (Unwin 1977).
D. R. *Poulter*, (ed.), *The Design of Gas-Cooled Graphite-Moderated Reactors* (Oxford University Press 1963).
Robin Russell *Jones* and Richard *Southwood* (eds), *Radiation and Health: The Biological Effects of Low Level Exposure to Ionizing Radiation* (Wiley 1987).
John *Simpson, The Independent Nuclear State* (Macmillan 1983, 1986).
Stuart *Sinclair, Windscale: Problems of Construction and Maintenance* (George Newnes 1960).
John *Valentine, Atomic Crossroads: Before and After Sizewell* (Merlin 1985).
The Uranium Institute, Understanding Chernobyl (1986).
Spencer R. *Weart, Nuclear Fear: A History of Images* (Harvard University Press 1988).
Roger *Williams, The Nuclear Power Decisions* (Croom Helm 1980).
T. I. *Williams, A History of Technology*, vol. VI (Oxford University Press 1978). Contains chapters on atomic energy by Lord Hinton.

Command papers

Cmd 9389, *A Programme for Nuclear Power* (HMSO 1955).
Cmnd 302, *Accident at Windscale No.1 Pile on 10 October 1957* (HMSO November 1957).
Cmnd 338, *Report of the Committee appointed by the Prime Minister to examine the*

Organisation of Certain Parts of the UKAEA (HMSO December 1957).

Cmnd 342, *Report of the Committee appointed by the Prime Minister to examine the Organisation for Control of Health and Safety in the UKAEA* (HMSO January 1958).

Cmnd 471, *Final Report of the Committee appointed by the Prime Minister to make a Technical Evaluation of Information Relating to the Design and Operation of the Windscale Piles, and to Review the Factors Involved in the Controlled Release of Wigner Energy* (HMSO July 1958).

Cmnd 1225, *The Hazards to Man of Nuclear and Allied Radiations: A Second Report to the Medical Research Council* (HMSO December 1960). Contains appendix on the 1957 Windscale accident.

UKAEA Annual Reports

First, second, third and fourth Annual Reports for the years 1954/55, 1955/56, 1956/57 and 1957/58.

Scientific Reports and Articles

H. M. *ApSimon*, D. *Earnshaw*, A. J. H. *Goddard* and J. *Wrigley*, 'Trajectory analysis with particular reference to the 1957 Windscale release' (Imperial College of Science and Technology, Nuclear Power Section. London, 1977).

K. F. *Baverstock* and J. *Vennart*, 'Emergency reference levels for reactor accidents: a re-examination of the Windscale reactor accident', *Health Physics*, 30 (1976), pp. 339–44.

J. R. *Beattie*, 'An assessment of environmental hazard from fission product releases', Culcheth, UKAEA, AHSB(S)R64 (1963).

H. W. *Bertini et al.*, 'Descriptions of selected accidents that have occurred at nuclear reactor facilities', ORNL/NSIC-176 NSIC.

Sir Douglas *Black*, *Investigation of the possible increased incidence of cancer in West Cumbria, Report of the independent advisory group* (HMSO 1984).

Joh *Blok*, R. H. *Dekker* and C. J. H. *Lock*, 'Increased atmospheric radioactivity in the Netherlands after the Windscale accident', *Applied Science Research*, 7 (1958), p. 150.

BNFL, Incidents at Sellafield involving abnormal releases of activity to the environment (1952-83). (SDB 239/W1) (1983).

M. *Bobrow*, *The Implications of the New Data on the Releases from Sellafield in the 1950s for the Conclusions of the Report on the Investigation of the Possible Increased Incidence of Cancer in West Cumbria: First Report*, COMARE (HMSO 1986).

D. V. *Booker*, 'Caesium 137 in dried milk', *Nature Lond.*, CLXXXIII, 4666, (April 4, 1959) 921-25

D. V. *Booker*, 'Caesium 137 in soil in the Windscale area', Harwell, AERE report R-40240 (1962).

D. V. *Booker*, 'Physical measurements of activity in samples from Windscale', AERE HP/R 2607 (1958).

British Medical Journal, 'Strontium 90 at Windscale', 27 August 1960 pp. 658–9.

F. J. *Bryant*, G. S. *Spicer*, A. C. *Chamberlain*, A. *Morgan* and W. L. *Templeton*, 'Radiostrontium in soil, grass, vegetables and milk from seven farms in the Windscale area', AERE HP/R 2636 (1958).

Bibliography

P. R. J. *Burch*, 'Measurements at Leeds following the Windscale reactor accident, Iodine 131 in human thyroids and iodine 131 and caesium in milk', *Nature Lond.*, 183 515 (1959).

A. C. *Chamberlain*, 'Comparisons of the emissions in the Windscale and Chernobyl accidents', AERE M3568 (1987).

A. C. *Chamberlain*, 'Deposition of iodine 131 in northern England in October 1957', *Journal of the Royal Meteorological Society*, 85 (1959), pp. 351–61.

A. C. *Chamberlain*, 'Dispersion of activity from chimney stacks', AERE M 409, May 1959 (declassified 1960).

A. C. *Chamberlain*, 'Emission of fission products and other activities during the accident to Windscale pile no.1 in October 1957', Harwell, UKAEA, AERE M3194 (1981, declassified 1983).

A. C. *Chamberlain*, 'Environmental impact of particles emitted from Windscale piles, 1954-7', AERE R 12163 (April 1986).

A. C. *Chamberlain*, 'Relation between measurements of deposited activity after the Windscale accident of October 1957', AERE HP/R 2606 (1958).

A. C. *Chamberlain* and H. J. *Dunster* 'Deposition of radioactivity in northwest England from the accident at Windscale', *Nature Lond.*, CLXXXII, 4636 (6 Sept. 1958), pp. 629–30.

R. H. *Clarke*, 'An analysis of the 1957 Windscale accident using the WEERIE code', *Annals of Nuclear Science and Engineering*, 1 (1974), pp. 73–82.

R. H. *Clarke*, 'A re-examination of the 1957 Windscale reactor accident', Proceedings of the 5th Annual Meeting of the Health Physics Society, July 1980, paper P/203; *Health Physics*, 39, 1056 (abstract only) (1980).

R. H. *Clarke*, 'Current radiation risk estimates and implications for the health consequences of the Windscale, TMI and Chernobyl accidents', UKAEA Conference on Medical Response to Effects of Ionising Radiation; London 28–30 June 1989.

R. H. *Clarke*, 'The 1957 Windscale accident revisited', International Conference on the Medical Basis for Radiation Accident Preparedness – II Clinical Experience and Follow-up since 1979: Oak Ridge, Tennessee, 20–22 October 1988.

A. *Cottrell*, 'Annealing a nuclear reactor: an adventure in solid-state engineering', *Journal of Nuclear Materials*, 100 (1981), pp. 64–6.

J. *Crabtree*, 'The travel and diffusion of radioactive material emitted during the Windscale accident', *Journal of the Royal Meteorological Society*, 85 (1959), pp. 362–70.

M. J. *Crick*, 'The collective dose from the 1957 Windscale fire (UK population)', NRPB Chilton, *Radiological Protection Bulletin*, 46 (May 1982), pp. 13–17.

M. J. *Crick* and G. S. *Linsley*, 'An assessment of the radiological impact of the Windscale reactor fire, October 1957', Chilton NRPB R 135 (1982).

M. J. *Crick* and G. S. *Linsley*, 'An assessment of the radiological impact of the Windscale reactor fire, October 1957', NRPB R 135 addendum (September 1983).

R. N. *Crooks*, K. M., *Glover*, J. W., *Haynes*, R. G. *Osmond* and F. J. G. *Rogers*, 'Alpha activity on air filter samples collected after the Windscale incident', Harwell, UKAEA, AERE-R 2952 (1959).

E. A. C. *Crouch* and I. G. *Swainbank*, 'Radiochemical and physical examination of the debris from the Windscale accident', Harwell UKAEA, AERE-C/R2589 (1958, declassified 1982).

A. *Cruickshank*, 'Re-assessing the dose from the Windscale fire', *Nuclear Engineering International*, 28, 339 (Apr. 1983), pp. 15–16.

H. J., *Dunster*, H. *Howells* and W. L. *Templeton*, 'District surveys following the Windscale accident, October 1957', *Proceedings of the 2nd UN Conference on the Peaceful Uses of Atomic Energy*, 18 (1958), pp. 296–308.

F. B., *Ellis*, H. *Howells* and W. L. *Templeton*, 'Deposition of strontium 89 and 90 on agricultural land and its entry into milk after the reactor accident at Windscale', AERE AHSB (RP) R2 (1960).

D. H. *Ehhalt* and A. E. *Bainbridge*, 'A peak in the tritium content of atmospheric hydrogen following the accident at Windscale', *Nature Lond.*, 209 (1966), pp. 903–4.

F. R. *Farmer* and J. R. *Beattie*, 'Nuclear power reactors and the evaluation of population hazards', Advances in Science and Technology, vol. 9 pp. 2–72 (Academic Press 1976).

G. *Gandusio* and S. *Polezzo*, 'The accident at Windscale and the radiation damage to graphite', *Energia nucleare* (Milan), 5 (May 1958), pp. 289–96.

R. J. *Garner*, 'The assessment of the quantity of fission products likely to be found in milk in the event of aerial contamination of agricultural land', *Nature Lond.*, CLXXXVI (1960), p. 1063.

John *Gittus* (ed.), 'The Chernobyl accident and its consequences' (UKAEA 1988).

R. *Herbert*, 'The day the reactor caught fire', *New Scientist*, 96, 1327 (14 Oct. 1982), pp. 84–7.

C. R. Hill 'Polonium-210 in man', *Nature Lond.*, CCVIII (1965), pp. 423–8.

H. *Howells*, A. E. *Ross* and R. *Gausden*, 'The release of oxide from irradiated uranium in the Windscale area since October 1955', UKAEA report IGO TM/W 036 (1957).

D. *Jakeman*, 'Notes on the level of radioactive contamination in the Sellafield area arising from discharges in the early 1950s', UKAEA report AEEW-R2104 (July 1986).

J. M. *Jones* and A. L. *Adams*, 'The Windscale piles – past, present and future', *Atom*, 374 (Dec. 1987), pp. 14–17.

G. *Kimber* and A. *Booth*, 'Radioactivity in plant material collected after the Windscale no.1 pile accident', *Nature Lond.*, CLXXXI (1958), p. 1391.

G. S. *Linsley*, 'the 1957 Windscale fire revisited', NRPB Chilton, *Radiological Protection Bulletin* (Nov. 1983), pp. 18–20.

G. S., *Linsley*, J. R. *Dionian*, J. R. *Simmonds* and J. *Burges*, 'Assessment of radiation exposure to members of the public in west Cumbria as a result of the discharges from BNFL Sellafield', NRPB Chilton R170 (1984).

J. F. *Loutit*, W. G. *Marley* and R. S. *Russell*, 'The Nuclear Reactor Accident at Windscale- October 1957: Environmental Aspects'. Appendix H in 'Hazards to man of nuclear and allied radiations', 2nd report to the MRC, Cmnd 1225 (1960), pp. 129–39.

MAFF 'Cattle illnesses in the Windscale area since the accident on 10 October 1957', ARC 97/58 (Jan. 1958).

W. G. *Marley* and T. M. *Fry*, 'Radiological hazards from an escape of fission products and the implications in power reactor location', UN Conference on Peaceful Uses of Atomic Energy (1955), Paper A/Conf.8.

G. *Maycock* and J. *Vennart*, 'Iodine-131 in human thyroids following the Windscale reactor accident. *Nature Lond.*, CLXXXII, 4649 (6 Dec. 1958) pp. 1545–47.

I. A. *Mossop*, 'Filtration of the gaseous effluent of an air-cooled reactor', *British Chemical Engineering* 5 (June 1960), pp. 420–6.

Nature, 'Resurrecting a nuclear accident', 302, 5905, 7 March 1983, p. 207.

G. H. *Palmer*, 'Environmental survey around an atomic energy site', HP/R2742 (1958, declassified 1986).

D. H. *Peirson*, R. S. *Cambray*, P. A. *Cawse*, J. D. *Eakins* and N. J. *Pattenden* 'Environmental radioactivity in Cumbria', *Nature Lond.*, 300 4 November (1982), pp. 27–31.

H. A. *Robertson* and I. R. *Falconer*, 'Accumulation of radio-iodine in thyroid glands subsequent to nuclear weapons tests and the accident at Windscale', *Nature Lond.*, CLXXXIV (28 Nov. 1959), p. 1699.

T. N. *Rutherford*, E. *Fanthorpe* and P. B. *Woods*, 'The Windscale piles – situation April 1961' (UKAEA IG report), AB62/71.

R. *Scott Russell*, R. P. *Martin* and G. *Wortley* 'An assessment of hazards resulting from the ingestion of fallout by grazing animals', AERE ARC/RBC 5 (1956).

B. S. *Smith*, 'Some lessons from the Windscale accident', Health physics in nuclear installations symposium (25–28 May 1959), pp. 395–401.

B. S. *Smith*, B. T. *Taylor* and C. O. *Peabody*, 'Some estimates of the scale of the Windscale accident based upon the noble gases released', AERE HP/M 133 (1958, declassified 1984) UKAEA.

F. W. *Spiers*, 'Measurements at Leeds following the Windscale accident. Increase in background gamma radiation and its correlation with I-131 in milk', *Nature*, 4660 (1959) pp. 517–19. *Nature Lond.*, CLXXXIII, 4660 (February 21, 1959) pp. 515–517.

F. W. *Spiers*, 'Some measurements of background gamma radiation in Leeds during 1955–59', *Nature Lond.*, 184 (28 Nov. 1959), p. 1680. CLXXXIV, 4700 (November 28, 1959) pp. 1680–82.

J. W. *Stather*, R. H. *Clarke* and K. P. *Duncan*, 'The risk of childhood leukaemia near nuclear installations', NRPB- R215 (1988).

J. W. *Stather*, J. *Dionian*, J. *Brown*, T. P. *Fell*, and C. R. *Muirhead*, 'The risk of leukaemia and other cancers in Seascale from radiation exposure', NRPB R171 addendum (April 1986).

J. W. *Stather* A. D. *Wrixon* and J. R. *Simmonds*, 'the risk of leukaemia and other cancers in Seascale from radiation exposure', NRPB R 171 (1984).

N. G. *Stewart*, and R. N. *Crooks*, 'Long-range travel of radioactive cloud from the accident at Windscale', *Nature Lond.*, CLXXXII 4636 (6 September 1958) pp. 627–28.

N. G. *Stewart*, R. N. *Crooks* and E. M. R. *Fisher*, 'Measurements of the radioactivity of the cloud from the accident at Windscale: data submitted to the IGY', Harwell UKAEA, AERE-M857 (1961, declassified 1962).

P. J. *Taylor*, *The Windscale Fire, October 1957*, report for the Union of Concerned Scientists, Cambridge, Mass. (PERG RR-7 1981).

D. *Williams*, R. S. *Cambray* and S. C. *Maskell*, 'An airborne radiometric survey

of the Windscale area, October 19–22 1957', EL/R2438 UKAEA (1958, declassified 1984).
D. *Williams*, R. S. *Cambray* and S. C. *Maskell*, 'An airborne radiometric survey of the Windscale area, October 19–22 1957', Harwell UKAEA, AERE-R2890 (1959).

Developments since 1995

Brian Cathcart

The fifteen years which have elapsed since the first publication of this book have seen no major developments that alter our understanding of the story of the fire, but work has gone on in the fields of decommissioning and the study of health effects.

The 1990s saw completion of a first phase of decommissioning, involving the clearing of fire damage and the installation of safety equipment around the core, which still in 2007 contains about 10 per cent of its fuel. Because of the difficulty of physically examining the contents, very pessimistic risk assumptions had to be made, and it appeared that further progress in safe dismantling would require the use of extremely complicated and costly techniques. But in 2002 UKAEA began using advanced computer technology to simulate the state of the reactor and a series of studies ranging over the most extreme scenarios led to the conclusion that the fire could not start again and the remaining fuel could not go critical. This finding made possible a new approach which has been approved by the contracting body, the Nuclear Decommissioning Authority, and by the Health and Safety Executive, and at the time of writing work is well advanced on the development of specialist equipment. The plan is to dismantle the reactor, starting at the top and working down, using a range of lightweight, high-payload robotic arms to remove the damaged fuel, graphite core, activated metals and concrete. The process is due to begin in 2008 and is expected to take about fifteen years.

So far as the health effects of the fire are concerned, monitoring has continued, largely under the auspices of the National Radiological Protection Board (which was absorbed into the Health Protection Agency in 2005). It remains the case that no health effect has been observed which can be attributed to the Windscale accident. A study* in 2000, carried out by Westlakes Scientific Consulting Ltd, looked at the group who might be thought to have been most at risk, the 470 male Windscale employees involved in fighting the fire and the clean-up operations that followed. It found no measureable effect on their mortality or cancer morbidity experience.

---

* D. *McGeoghegan* and K. *Binks*, 'Mortality and cancer registration experience of the Sellafield employees known to have been involved in the 1957 Windscale accident', *Journal of Radiological Protection* 20 (2000), pp. 261–74.

Index

accident inquiries, rules and
procedures, 64–5, 87
accident at Windscale
consequences, 155–8
financial cost, 157
impact on production, 75–6, 158
organisational results, 103, 107–10,
156–8
technical lessons, 91, 123, 134,
155–7
AEA (Atomic Energy Authority)
Board, 81–2
creation of, 19–20, 102
resources and commitments, 26–9
Aesop's frog, 159
AEX (Atomic Energy Executive), 66,
75, 80, 81
AGR, Windscale, 24
AHSB (Authority Health and Safety
Branch), 107–8, 110, 156
Radiological Protection Division,
110
Safeguards Division, 110
Aldermaston, 26, 30, 73
Allaun, Frank, MP, 69
Allen, Arnold, 52, 57, 58, 65
AM and AM cartridges, 26, 30, 31, 44,
67, 76, 114–15, 122, 128–34,
172–4
see also tritium, lithium-magnesium
Anderson, Frank, MP, 62, 67, 81
Anglo–American atomic relations,
3–6, 18, 83–5, 96, 158
Anglo–American information
exchanges, 3, 13, 33–4, 89, 90, 95,
127, 139
ARC (Agricultural Research Council),
55, 63, 70, 73, 179

Baverstock, K., 144
BCDG (burst cartridge detection gear),
13, 17, 18, 34, 35, 36, 48, 51, 88,
120, 130

Beattie, J. R., 143
Belgium, 95, 96
Bell, J. C., 42
BEPO reactor, Harwell, 1, 8, 10, 96,
111
biological monitoring, 36, 53, 72–3
bismuth oxide cartridges, 29, 44, 174
see also polonium-210
Black report, on incidence of cancer in
West Cumbria, 148
blowers, booster fans and shutdown
fans, 12, 16, 32, 45
BNFL, 110, 159
Bowen, J. C., 128–9
Britain's great-power status, 6
Brook, Sir Norman, 82, 93
Brookhaven, USA, 13, 95
Brown, George, MP, 69, 81
Brundrett, Sir Frederick, 81
burst cartridges, 13, 34–37, 169
see also BCDG

caesium-137, 63 et passim
Calder Hall 11, 21, 23, 39, 40, 51, 54,
71, 75, 88, 92, 159
see also Magnox
Calder Operations school, 23, 109
Canada, atomic research
establishment, 4, 8, 10, 12
Capenhurst, 3, 68, 74
carbon dioxide, 49, 51
Carling, Sir Ernest Rock, 63
cartridges, 114–15, 131–2, 169–74
see also AM, bismuth oxide,
uranium fuel elements
cause of fire, 78, 120, 125–35
CEA (Central Electricity Authority)
and CEGB (Central Electricity
Generating Board), 41, 75, 82, 83
Chamberlain, Dr A. C., 145, 148
Chapelcross, 11, 21, 39, 40, 88, 92
see also Magnox, PIPPAs
Chernobyl accident 1986, 151–2

Index 233

Chicago conference on reactor safety 1956, 34, 90
chief safety officer, Risley, 56, 61, 106
Chief Constable of Cumberland, 50, 52
'Christmas trees', 13, 17
  see also BCDG
Churchill, Sir Winston, 3, 4, 6, 18, 19, 25
Citrine, Lord, 83
civil nuclear power programme, 11, 21–4, 39, 82, 88, 159
  see also Magnox, Trend report
Clarke, R. H,. 143, 152–3
Cockcroft, Sir John, 4, 5, 6, 13, 14, 20, 35, 66, 68, 81, 88, 90, 95, 105, 106, 140
'Cockcroft's follies', 13, 14, 39
  see also pile stack filters
Cold War, 6
collective responsibility for accident, 86–7, 215
COMARE report on cancer incidence in West Cumbria 1984, 148–9
control rods, 16, 45, 46
consultants, radiological, 36, 58, 61–2, 73, 137
Cook, sir William, 103, 104
coolant questions, 9–11, 16
Cottrell, sir Alan, 89, 91, 92
Cumberland County Council, 57, 71
  see also Chief Constable

Daily Express, 94
Daily Telegraph, 122–3
Davey, H. G. (Gethin), 31, 38, 39, 40, 41, 44, 48, 49, 50, 51, 52, 55, 58, 69, 90, 127, 133, 158, 162
decommissioning, 159–60
defence requirements paramount, 27, 102, 154, 158
  see also timetables
Diamond, Professor Jack, 66, 68, 127
Dounreay, 24, 40, 101
(10) Downing Street, 62, 65, 67, 85
dual purpose reactors, see PIPPAs
Dunster, H. J., 56, 57, 72, 110, 137, 138
Dunworth, J., 90

Economist, The, 67, 93, 94
Eisenhower, President, 85
emergency planning, 50, 59, 175–7
enriched uranium fuel, 8, 31
environmental contamination, 37, 53–7, 64, 96–7, 140, 141
  see also pre-accident emissions, radioactive releases
environmental surveys, 35–7, 54, 71–3
Europe, information exchanges, 95–7
evacuation plans, 50, 52, 175
expending commitments, 28–9
  see also overload, staff shortages

Failla, G., 13, 89
Fair, D. R. R., 127, 133
fallout, 2, 56, 63, 141
Farmer, F. R., 56, 57, 58, 61, 106, 108, 110
farmers, 213–4
  see also NFU, Wallbank
fast reactors, 8, 24
filters, see pile stacks
'fin-clipping', 15
fire-break, 49, 50
fire brigade, 50
fire hazards, 9, 13
fire officer, 50
fission products, 8, 54, 63, 72, 112, 138, 143–4, 184–6 (only the most important fission products are indexed individually)
Fleck, Sir Alexander (later Lord), 26, 68, 80, 81, 82, 93, 98, 104
Fleck Committee
  on Health and Safety, 104–10
  on Organisation, 98–104, 155
Fleck Technical Evaluation Committee, 110–17
Fletcher, P. T., 68
food chain, 53, 54, 63
France, 96
Fry, T. M., 54. 90
fuel elements, see uranium fuel elements, cartridges
fuel loading machine, 11–12

Gaitskell, Hugh, MP, 67, 80
Gardner, Professor Martin, 150

Gausden, Ron, 40, 42, 44, 47, 48, 49, 119
Geneva conference on peaceful uses of atomic energy (1958), 138
GLEEP reactor, Harwell, 8
Goodwin, Vic, 42
Graham, Dr T. ('Thos'), 55, 56
graphite problems, 12–13, 34, 89–92, 113, 121–2, 126–7, 132–3
Greenough, G., 91

Hailsham, Lord, 67, 69, 80, 87
Handford; USA, 4, 8, 9, 11, 12, 33, 95
Harwell, 4, 7, 8, 10, 12, 13, 15, 33, 56, 89, 97 *et passim*
Harwell Reactor School, 23
Harwood, Sir Edmund, 58, 61
health and safety, 87, 104–10
health and safety committees, 105–7
health and safety organisation, 78–9, 104–10, 155–7
health impact of accident, 64, 78–9, 87, 136–50, 152–3, 187
health physics, 53, 54, 95, 105
Hill, Dr (now Sir) John, 128, 129–31, 133–4
Himsworth, Sir Harold, 63, 64, 80, 87, 137, 138, 139
Himsworth Committee on 'Hazards to Man of Nuclear and Allied Radiations', 63, 64, 73, 139–43
Hinton, Sir Christopher (later Lord), 4, 5, 6, 9, 10, 11, 12, 15, 17, 20, 26, 28, 35, 38, 41, 64, 83, 87, 101, 104, 109, 123, 155, 156
Howells, Huw, 36, 40, 48, 53–9, 72, 138
Hughes, Tom, 40, 48, 53
Hurricane A-bomb test, 1952, 7, 18, 19, 21
hydrogen bomb, 25–6, 30, 75, 84

ICRP (International Commission on Radiological Protection), 55, 57, 108, 139, 142, 146, 152
IG (Industrial Group, AEA), 98–104 *et passim*
instrumentation, 16–17, 81, 113–4, 119 *see also* thermocouples

iodine-131, 54–5, 57, 60–1, 62, 63, 71, 112, 137, 140, 145, 178–83
IPCS (Institution of Professional Civil Servants), 94, 115, 117–21, 128, 135
Ireland, G. D., 40
irradiation facilities, 30, 75, 90
Italy, 95

Jakeman, Dr D., 148
Japan, 95
Jenkinson, Peter, 42
Jukes, J. A., 81, 88

Kay, Professor J. M., 66, 68
Kearton, C. F. (later Lord), 104
Kronberger, Dr Hans, 67, 91

Latina power station, 95
Leslie, Dr Frank, 62
leukaemia and other cancers, studies of incidence, 64, 147–50
see also Black, COMARE, Gardner
lithium-magnesium alloy, 30, 78
see also AM cartridges
LM, *see* polonium-210, bismuth oxide cartridges
Loutit, Dr. J. F., 63

McLean, Dr A. S., 55, 56, 57, 58, 61, 68, 104, 106, 108, 110, 137
McMahon Act (USA), 1946, 5, 84, 85
Macmillan, Harold, *see* Prime Minister
Magnox alloy, 21
Magnox fuel, 30
Magnox reactors, 11, 23–4, 81, 88, 92, 134, 158
see also Calder Hall, Chapelcross, civil power programme, PIPPAs
Manchester Guardian, 62
Manhattan Project, 3, 8, 9
Marley, Dr W. G., 9, 54, 57, 58, 63, 108, 140
Mayneord, Professor W. V., 37
milk, 37, 54, 55, 57, 64, 71
see also iodine-131, strontium-90
milk ban, 57, 61, 72–3, 137, 140, 146, 153
Milk Marketing Board, 58, 60, 70

Mills, Lord, 83
miners, 70, 213
   see also NUM
Ministry of Agriculture, Fisheries and Food (MAFF), 35, 37, 43, 58, 61, 64, 70, 179
Ministry of Defence, 80–1, 83
Ministry of Fuel and Power, 109
Ministry of Health, 61
Ministry of Housing and Local Government (MHLG), 35, 61
Ministry of Supply (MOS), 5, 19, 21, 29
Ministry of Works, 29
Mitchell, Professor J. S., 57
'monuments of ignorance', 17, 123
MRC (Medical Research Council), 37, 63, 80, 87, 138, 139, 141, 181–3
   see also Himsworth, Himsworth Committee

New Scientist, 93
NFU (National Farmers' Union), 69, 70
   see also farmers
NII (Nuclear Installations Inspectorate), 92, 109, 156
NRPB (National Radiological Protection Board), 109–10, 146–7, 152, 156
Nuclear Installations (Licensing and Insurance) Act 1959, 109
'nuclear umbrella', 6
nuclear weapons and weapon tests, 6, 7, 18, 19, 25, 66, 84, 158
NUM (National Union of Mineworkers), 70
   see also miners

Oak Ridge, USA, 13, 14
Operations Branch, Industrial Group, 30, 98, 100–1 *et passim*
overload, 26–9, 124, 154–5
Owen, Sir Leonard, 20, 40, 41, 43, 58, 61, 66, 68, 88, 91, 95, 101, 102, 103, 104, 106, 123, 155, 158, 162

Parliament, 5, 19, 25–6, 27, 68–9, 80, 82, 93

Peart, Fred, MP, 69, 81
Peirson, David, 77, 86–7, 118
Penney, Sir William (later Lord), 5, 6, 7, 20, 26, 66, 77, 81, 86–7, 91, 98, 104, 106
Penney Inquiry and Report, 67, 77–85, 115, 126, 138, 189–204
Perrott, Sir Donald, 117–8, 120
Phillips, J. L., 33
piles, plutonium production, 10–12, 15–16, 29–31, 89, 159–60
   see also PIPPAs
pile no. 1, Windscale, 14–15, 17, 18, 32, 36, 42–52, 74, 116–117
pile no. 2, Windscale, 17, 18, 32, 35, 60, 75, 80, 116–17, 121–3, 132–3
pile stacks and pile stack filters, 13–14, 37–39, 43, 54, 60, 74, 111–13, 145
PIPPAs (dual purpose reactors), 21, 28
   see also Magnox reactors
Plowden, Sir Edwin (now Lord), 20, 35, 37, 43, 58, 60, 61, 66, 67, 82, 83, 84, 86, 93, 94, 95, 104, 105, 111, 123, 134, 158, 215
plutonium, 2, 3, 4, 5, 6, 7, 17, 18, 21, 24, 28, 75
Pochin, Dr (later Sir Edward), 180
polonium-210, 29–30, 76, 97, 138, 140, 146, 147, 152–3
Porton (chemical defence research establishment), 38
pre-accident radioactive emissions, 35–7, 64, 141, 148–9
   see also environmental contamination, radioactive releases
press and public relations, 58, 69–70, 93, 94, 213
Prime Minister, 37, 43, 60, 62, 65, 66, 68, 80, 82, 83–5, 93, 94, 117, 123
priorities, 29, 99, 155
Production Division, Ministry of Supply, 6–7

Queen Elizabeth II, 71
radiation, 221
radiation standards, limits and risk coefficient, 2, 55, 57, 63, 80, 141, 142, 150, 152, 178, 183

radioactive releases from accident, 47, 49, 112, 138, 140, 143, 145–7, 184–6
radioactive releases, pre-accident, 35–7, 64, 141, 148–9
radioisotopes, 30, 75
Rennie, C. A., 13
R & D Branch (Rilsey and Windscale), 34, 40, 42, 49
research, deficiencies in, 118, 156–7
risk-taking, 102, 158
Risley safety organisation, 4, 6, 25, 34, 55, 58–9, 106–7
Robertson, Ian, 40, 42, 45, 46
Ross, K. B., 43, 49, 50, 51, 52, 56, 58, 68
Rotherham, L., 68
Rothschild, Lord, 70
Russell, Dr. R. Scott, 55, 61

Salisbury Lord, 65
Schonland, Sir Basil, 66, 95, 158, 162
Seascale, 53, 62, 149–50
Sallafield, 11, 176
Sheard, H., 89
siting problems, 9–11
Sputnik, 84–5
stacks and stack filters, see pile stacks and pile stack filters
staff shortages and staffing problems, 20, 26–9, 40, 48, 98–103, 108, 155, 211
Stewart, J. C. C., 68, 127, 128, 133
Stewart, N. G., 140
Strath, Sir William, 115
strontium-90, 37, 54, 57, 63, 64, 73, 141

Taylor, Peter, 146
TEC (Technical Evaluation Committee), 68, 110–17
Teller, Edward, 13, 89
Templeton, W. L., 72, 138
thermocouples, 32, 33, 44, 45, 46, 47, 113–14, 130
Three Mile Island accident 1979, 71, 144, 214
thyroid doses, 55, 73, 139, 141

timetables, inexorable, 7, 27, 154–5
Tokai-Mura power station, 95
Trend report on civil nuclear power programme, 1954, 22–3
tritium, 26, 30, 75–6, 97
 see also AM
Tuohy, Tom, 15, 17, 39, 40, 49, 50, 51, 52
TX (Technical Executive Committee), 68, 126

Underwood, Eric, 58
UNSCEAR (United Nations Scientific Committee on Effects of Atomic Radiation), 108, 152
uranium, 1, 2, 5, 169–71
uranium fuel elements, 8, 13, 14, 16, 34, 169–71
 see also cartridges, fuel elements
USA, see Anglo–American relations, Anglo–American information exchanges, Brookhaven, Chicago conference, Hanford, Manhattan Project, McMahon Act, nuclear weapons
USAEC (US Atomic Energy Commission), 58, 95
USSR, see Cold War, nuclear weapons, Sputnik

Veale report on health and safety training, 110
Vennart, J., 144

Wallbank, Mr, 70, 71
 see also NFU
water, used to extinguish fire, 51–2
Waverley report (on setting up the AEA), 19–20
Whitehaven, 58, 177
Wigner effects, 12–13, 31–4, 42–8, 88, 167, 168
 see also graphite
wind pattern, 53
Windeyer, Sir Brian, 104
Windscale Technical Committee, 36, 39, 42